SECURITY REQUIREMENTS ENGINEERING

The MIT Press Information Systems Series

Michael Papazoglou, Eric Yu, Florian Matthes

SECURITY REQUIREMENTS ENGINEERING
Designing Secure Socio-Technical Systems

Fabiano Dalpiaz
Elda Paja
Paolo Giorgini

The MIT Press
Cambridge, Massachusetts
London, England

This book was set by the authors using LaTeX. Printed and bound in the United States of America.

Library of Congress Cataloging-in-Publication Data

Names: Dalpiaz, Fabiano, author. | Paja, Elda, author. | Giorgini, Paolo, author.
Title: Security requirements engineering : designing secure socio-technical systems / Fabiano Dalpiaz, Elda Paja, and Paolo Giorgini.
Description: Cambridge, MA : MIT Press, [2015] | Series: Information systems | Includes bibliographical references and index.
Identifiers: LCCN 2015038374 | ISBN 9780262034210 (hardcover : alk. paper)
Subjects: LCSH: Systems software—Design and construction. | Sociotechnical systems—Computer-aided design. | Requirements engineering—Data processing. | Technology—Social aspects. | System analysis—Data processing.
Classification: LCC QA76.76.S95 D35 2015 | DDC 005.4/3—dc23 LC record available at http://lccn.loc.gov/2015038374

10   9   8   7   6   5   4   3   2   1

To our families.

Contents

List of Figures

List of Tables

Preface

Security requirements engineering has received increasing attention over the past decade: security is a crucial quality attribute of any software system, and building secure software requires an analysis of security that starts from the very early development phases. This task is especially challenging because designers have to study the bigger picture, which includes not only the software under design, but also the humans operating and interacting with it, the organizations involved, and so on. Analyzing this bigger picture means designing a secure *socio-technical system*, rather than a merely technical system.

Socio-technical systems are not just futuristic. Even though research efforts into their design are still in their infancy, socio-technical systems have been a reality for quite some time. Their societal relevance and diffusion is demonstrated by many examples, such as healthcare systems, e-commerce, air traffic management control, smart cities, smart homes, and so forth.

The socio-technical perspective poses new challenges for understanding and specifying the security requirements for the system-to-be. These challenges are related to the complexity of these systems, the dynamics of the behaviors and interactions of the underlying subsystems, the autonomous and heterogeneous nature of those subsystems, and the lack of a controlling authority imposing or enforcing requirements on them.

This book proposes the Socio-Technical Security method (STS) for engineering the security requirements of socio-technical systems. STS provides guidance for a systematic security requirements engineering process. The method is model-driven, and comes with the Socio-Technical Security modeling language (STS-ml), which provides the necessary concepts and relationships to express security requirements for a socio-technical system.

The book provides a comprehensive account of the STS method, STS-ml, and the supporting software tool (called STS-Tool), which facilitates modeling and reasoning activities supported by the method.

Audience. This first edition of the book addresses two target audiences: for an *academic audience*, it serves as a textbook to teach security (requirements) engineering; for *practitioners*, it is a reference for the early adoption of a state-of-the-art security requirements engineering method. The book reports on practical applications of the method in two industrial case studies. More efforts are needed, however, to foster industrial uptake, and this will be the aim of future editions of this book.

Acknowledgments. We wish to thank and acknowledge all those who provided us valuable feedback and comments throughout the process of completing this work. We owe special thanks to John Mylopoulos and Fabio Massacci (University of Trento) for their continuous and valuable feedback. For STS-Tool, we are also much indebted to the lead developer Mauro Poggianella and to Pierluigi Roberti. Many thanks to Alexander Borgida (Rutgers University) for the interesting discussions and valuable comments on the automated analysis framework. We are very grateful to the industrial partners of the EU Funded FP7 Project Aniketos, who have adopted the STS method, used STS-ml and STS-Tool since the beginning, and provided continuous feedback for their improvement. Above all, we are much indebted for their participation in the evaluation workshops. We are grateful to Stéphane Paul (Thales Research and Technology) and Per Håkon Meland (SINTEF) for their constructive criticism, and to Sandra Trösterer and Elke Beck (University of Salzburg) for all the help and support in organizing the evaluation workshops. We thank all the people at MIT Press who helped us throughout the process of writing and publishing this book, especially Marie Lufkin Lee, Katherine A. Almeida, and Kathleen Hensley. Finally, we thank the anonymous reviewers for their comments to improve the presentation of this book.

Book organization. The book is structured in five thematic parts, with nine chapters that present the context, introduce the approach, illustrate it on concrete scenarios, and compare it with alternative approaches.

Part I serves as an introduction to the book. *Chapter 1* presents the landscape of security requirements engineering, emphasizes the need of considering security from a socio-technical perspective when designing software systems, and describes the running example employed throughout the book. *Chapter 2* provides an overview of information and computer security, thereby introducing the necessary terminology for reading the remainder of the book.

Part II presents the STS-ml security requirements modeling language for socio-technical systems. *Chapter 3* introduces each of the modeling concepts supported by the language. This chapter is the reference to consult whenever unable to understand parts of an STS-ml model. *Chapter 4* combines the primitives of STS-ml into three different views (social, information, authorization), each of them representing a different perspective to be considered when conducting security requirements engineering.

Part III expands the view on STS-ml by explaining how to use it within a method for security requirements engineering. *Chapter 5* details automated reasoning techniques that an analyst would conduct to iteratively refine the model into a consistent one where no security requirement is violated. *Chapter 6* describes the STS method that provides guidelines for the creation and refinement of STS-ml models.

Part IV puts the proposed approach in practice. *Chapter 7* introduces STS-Tool, the software tool that accompanies STS-ml and is at the basis of the STS method. STS-Tool implements the framework presented in Part II and Part III. *Chapter 8* illustrates two applications of the method on scenarios from industrial case studies: the former concerns an online collaborative platform, the latter considers an e-Government system.

Part V concludes the book by comparing STS-ml and the STS method with alternative and complementary approaches in the area of security requirements engineering. This is done in *Chapter 9*, which also explains whether these techniques can be used in conjunction with STS or constitute an alternative.

How to read this book. The chapters of this book are best read in sequence. Some of the chapters and sections can be skipped or skimmed through, if the reader is already familiar with the matter at hand. However, we warn against relying solely on background knowledge, for some concepts are redefined in the book (especially in Part I); ignoring this advice may result in false assumptions leading to misunderstandings.

We strongly advise all readers to thoroughly read Part II, which details the STS modeling language. This is the core part of the book, and failing to understand these notions would hinder comprehending the remainder of the book. Our advice also applies to the reader with experience in goal-oriented requirements engineering, for some of the primitives in STS-ml significantly differ from mainstream frameworks such as i^* [69], Tropos [6], Secure Tropos [42], and SI* [20].

We recommend reading the automated reasoning techniques in Chapter 5; however, it is possible for the user of the STS-Tool to consult that chapter when necessary (e.g., when the tool detects errors in the model, and the user does not understand their meaning).

We also suggest reading the STS method presented in Chapter 6, which describes the recommended way of using STS-ml.

Part IV and Part V are optional reading. The description of STS-Tool in Chapter 7 should be seen as a concise overview of its capabilities, but it can be replaced by the user guide available on the tool's website (http://www.sts-tool.eu). The case studies in Chapter 8 can be consulted whenever the reader needs to clarify some aspects of STS. Chapter 9 is useful for the reader who is interested in comparing the approach to other works in security requirements engineering.

I INTRODUCTION

1 Security Requirements Engineering

The importance of computer and information security is before our eyes. We regularly come across news concerning security failures: stolen passwords and identities, systems infected by viruses and worms, websites made unavailable by (distributed) denial of service attacks, and many others. The cost of information security breaches keeps increasing, and the incurred financial losses are in the range of hundreds of thousands of dollars to millions of dollars per year, despite good signals such as increased investments in security [47, 48].

The design of secure software is an integral part of the security landscape. Many breaches occur because systems are designed and engineered with a poor understanding of security needs, as an *afterthought* [58]. In addition to the well-known vulnerabilities that open the doors to malicious users, these ill-designed systems exhibit less visible flaws, such as the sharing of unnecessary information with legitimate users, thereby exposing organizations to the most frequent cause of security incidents: their own employees [48].

Researchers in software engineering, and especially in requirements engineering, have acknowledged the importance of considering security early in the development process [11, 12]. These efforts fall within the field of *Security Requirements Engineering (SRE)*; the purpose of this chapter is to provide the reader with an overview of the field, and provide solid arguments that justify (i) considering security from the very inception of software engineering projects, and (ii) taking into account the larger socio-technical environment in which software systems operate.

This chapter reviews the origins of security requirements engineering (Section 1.1), illustrates how socio-technical systems are pervasive in our daily life (Section 1.2), explains the importance of considering security challenges from the socio-technical perspective (Section 1.3), and justifies the need for a novel approach (Section 1.4). Finally, the chapter introduces the running example concerning healthcare that is used throughout the book (Section 1.5).

1.1 The dawn of security requirements engineering

Security requirements engineering emerged in the early 2000s as a response to the substantial monetary expenditures required to cover the damages provoked by security breaches. Many researchers all over the world contributed to the establishment of this new discipline [12]; the purpose of this chapter is to present a concise history, in order for the reader to understand the background that led to the work described in this book.

One of the pillars of SRE is to promote *security by design*, that is, a development practice in which security issues are considered as early as possible [11, 12], from the requirements engineering phase, to inform the later phases such as design, implementation, and testing. The pioneering work by Devanbu and Stubblebine [11] paved the way for a large number of proposals for SRE [41], which is now well established within requirements engineering.

Privacy was long considered as part of security, being captured as confidentiality. Similarly to security by design, however, privacy by design has evolved to emphasize the need of adopting a proactive approach to protecting data, offering a broader view of privacy, including *privacy as control* and *privacy as practice* [4, 24].

It is worth observing that a significant number of approaches within SRE advocate the use of *modeling languages* to create models of the system to be designed in a secure manner. Model-driven security requirements engineering offers the advantages of precisely documenting and analyzing security requirements together with design requirements [1].

Early works in SRE considered security as a type of non-functional requirement [35, 68], introducing quality constraints under which the system must operate. This paradigm is adequate for a very high level analysis of security but offers only limited support for later stages of requirements engineering—which study assets, threats, authorizations, prohibitions, and so on—that guide the integration of security solutions into a software system design [19].

Another family of early approaches proposed to identify and specify security requirements in a system-oriented fashion [18, 25, 39, 40, 56, 62], where the system is treated as a monolithic entity while focusing strictly on technical mechanisms, and the only considered interactions are those among end users and the system. The main limitation of these approaches is that they fail to capture the social and organizational context in which the software system operates.

1.2 The era of socio-technical systems

The failure of large complex systems to meet their deadlines, costs, and stake-holder expectations are not, by and large, failures of technology. Rather, these projects fail because they do not recognize the social and organizational complexity of the environment in which the systems are deployed. The consequences of this are unstable requirements, poor systems design and user interfaces that are inefficient and ineffective.
—Gordon Baxter and Ian Sommerville [2]

Most of today's software systems are part of larger socio-technical systems [10, 15, 57], which include not only technical components (e.g., software), but also social ones such as humans and organizations. The subsystems within a socio-technical system are independent and weakly controllable by others. For example, the writing process of this book has defined a socio-technical system that includes the three authors, their respective employing organizations, the publisher, the reviewers, email clients and servers, typesetting programs, video-conferencing systems, collaborative modeling environments, and more.

The subsystems within a socio-technical system establish social dependencies for achieving objectives that they are not able to fulfill on their own, or that they prefer delegating to others. For example, the authors of this book depend on the publisher for printing copies of the book, on reviewers for providing constructive feedback, on video-conferencing systems for the remote discussion of the chapters, and so forth.

While depending on each other, the subsystems exchange a considerable amount of information. For instance, the authors transfer the drafts of this book to the editors, who, in turn, send these copies to the reviewers; also, the authors transfer their personal details to the publisher to establish a contract for publishing this book as part of the MIT Press Information Systems book series.

Socio-technical systems pervade our lives. Examples include prominent virtual communities wherein people interact through some form of social media; the e-commerce platforms that we use every day to buy and get items delivered at home; today's and tomorrow's smart cities and their components, such as smart homes; modern healthcare systems with their increasing reliance on electronic systems; (higher) education systems; and many others.

1.3   Security in socio-technical systems

The mantra of any good security engineer is: "Security is a not a product, but a process." It's more than designing strong cryptography into a system; it's designing the entire system such that all security measures, including cryptography, work together.
—Bruce Schneier [53]

The previous section explained how socio-technical systems go beyond the domain of software systems. This implies that the engineering of a secure socio-technical system requires considering the interplay of software systems with the humans and organizations that participate in the socio-technical system. In such systems, security issues are not just technical but also social, and they arise because of the *interactions* of participants with one another in the socio-technical system.

Maintaining security in socio-technical systems is especially hard because there is no central controlling authority to impose or enforce compliance of participants' behavior with what is expected from them. They might join or leave the system as they please. Thus, they might fail to comply with tasks required by other participants, leaving the latter vulnerable or prone to dealing with a dead end.

Several researchers in security requirements engineering have proposed approaches that analyze security from a broader socio-technical perspective. Many of these approaches rely on goal- and actor-oriented models—often variants of Eric Yu's *i** [66] framework—and represent a socio-technical system as a set of actors[1] that are *intentional* (they have goals) and *social*, as they interact with others to achieve their desired objectives. Some approaches [20, 36, 42] consider security by distinguishing among actors that own a service or a resource (assets), actors requiring a service/resource, and actors entitled to do any of these. These approaches define the baseline of the method presented in this book.

---

1 Actors represent system participants, also referred to as stakeholders.

1.4 On the need of a new approach

The reader may legitimately wonder why a new approach to security requirements engineering is needed. The answer comprises several reasons, which are briefly illustrated in this section, and lay the foundations for the principles of the proposed approach, which are thoroughly described in Section 3.1.

Operationalizability. Existing goal- and actor-oriented approaches to security requirements engineering are an adequate baseline, for they explicitly recognize the importance of social factors. They are mainly intended for the very early stages of requirements engineering, however, and employ modeling primitives that are difficult to operationalize into technical requirements for the system under design.

Real-world requirements. To gain acceptance by practitioners, a modeling language should employ modeling primitives that are well aligned with the way stakeholders would express their concerns regarding security. Most existing approaches use primitives that are too far from those that stakeholders and security analysts use in practice.

Adherence with standards. The practitioners' community has emphasized the need for adherence with the security principles followed by the information security community, such as the terminology and best practices that are listed in security standards [22, 27–29, 32, 46].

Expressing security needs. Any approach to (security) requirements engineering should keep track of who expresses a certain (security) need. This is particularly true in socio-technical systems, which explicitly account for the existence of multiple participants, each having a say in what (security) requirements they want others to comply with. While they represent multiple actors, existing approaches do not explicitly relate (security) needs with the entity that requests them.

Generic support for risk analysis. Security requirements engineering methods cannot overlook the security issues that arise because of malicious events threatening participants' assets. Most organizations, however, already employ a risk analysis method, and it would be impractical to impose one specific method. Therefore, this book leaves freedom to choose a preferred method to analyze risk and proposes using the obtained results to inform the identification of the security requirements for the system-to-be.

Clear representation of assets. In line with [19], effective approaches require distinguishing between different types of assets, thoroughly analyzing asset manipulation and exchange, and identifying the security needs of the asset owners. This differentiation is core to a clear understanding of *why* security is needed, *what* should be protected, and *what* security needs the stakeholders have, to then answer *how* these security needs can be satisfied. This distinction is only marginally considered by existing approaches.

Some of these limitations are well covered by approaches to SRE that are not goal- and actor-oriented. Those works serve as inspiration for the proposed method. Chapter 9 presents a comparison with a number of complementary and alternative and well-known methods to SRE, showing how they can be synergistically used with STS.

1.5   Running example: healthcare

Healthcare is a prime example of a socio-technical system: organizations (e.g., hospitals, laboratories) interact with one another and with humans (e.g., doctors, nurses, patients) to provide healthcare services, through the use of technical systems. For example, doctors use medical equipment to visit patients and prescribe medications; citizens provide their personal data to be hospitalized and receive healthcare; hospital staff enter patient data into a medical information system, and so forth.

This book illustrates the proposed approach employing a healthcare scenario, abstracted from the case of the Hong Kong Red Cross Blood Transfusion Center (for brevity Red Cross BTC), described in the following paragraphs. A detailed description of the case is publicly available at http://www5.ha.org.hk/rcbts/enindex.asp.

Alice is one of many donors who periodically donate blood through the Red Cross BTC. The center is responsible for collecting and examining the blood collected from the donors, and then distributing it to different hospitals. The Red Cross BTC is also responsible for determining the eligibility of potential donors. Elaborate test results, however, are performed at specialized laboratories. For instance, Alice undergoes infectious disease testing at ModernLabs.

Hospitals maintain health records for all their patients, and perform transfusions for certain patients. All information used and maintained throughout the transfusion procedure, including the type of operation, participating medical practitioners, and the reason for transfusion, is documented and stored in the database of each hospital. Hospitals rely on physicians to provide healthcare services to patients. Physicians of the hospital access this information to provide adequate medical health advice and treatments to patients.

Patient information and blood usage listings can be accessed by the Red Cross BTC to perform statistical analysis and auditing tasks, in order to provide accurate estimations on future blood consumption in different hospitals, and to make recommendations on blood usage in future medical cases.

The Hospital Authority sets the regulations concerning the privacy of patient records. The Red Cross BTC submits reports to the Hospital Authority, which has to keep patient privacy protected, in accordance with the regulations.

The successful operation of the Red Cross BTC socio-technical system depends on effective and efficient interaction among donors, physicians, laboratories, patients, hospitals, the Hospital Authority, and the Red Cross BTC itself. Each participant has its own expectations of security, which constrain the way they would like others to behave when it comes to the assets they want to protect. For example, donors allow the Red Cross BTC to use their data for approving them as donors and for any statistical analysis needed over the collected blood, but they do not want the involvement of third parties, even of research centers. Celebrities being hospitalized, in particular, want the information related to their conditions to remain confidential and not be made public.

2 An Overview of Computer and Information Security

Computer security is the discipline concerned with the protection of *assets* that are relevant to a computer or an information system. According to the *Oxford Dictionary of English*, an asset is "a useful or valuable thing or person." In computer security, many types of assets exist, including the hardware, software, information, processes, and people that are necessary for the operation of a computer or information systems.

Computer security is an especially challenging activity in organizations, which are complex systems that are defined by the interactions among individuals, processes, and technical systems. These interactions occur either verbally, via written documents, or through the use of the organization's information systems. Moreover, organizations are asset-rich environments, which contain confidential documents, client contacts, employee passwords, email messages, software programs, employees, computers, the business processes that define how the organization operates, and so forth.

In organizations, significant effort is put into securing the *informational assets*, because of the pivotal role that information systems play in organizations and the large amount and importance of information within modern organizations. Think, for instance, of the information that a hospital typically possesses about its patients and employees.

The key role of information in our daily lives justifies *information security* as a prominent, distinctive concept within the security realm. While this book does not focus on information security only, many of the security requirement types that our approach supports are closely related to information security.

This chapter provides a brief overview of computer and information security. In Section 2.1, we define a taxonomy of key security concepts. In Section 2.2, we explain how security issues arise by reviewing the terms of threat, risk, and vulnerability. In Section 2.3, we present an overview of the most fundamental security mechanisms to cope with existing risks.

The chapter is not intended as a replacement for security textbooks. The interested reader can find detailed information concerning the field in the books by Gollmann [22], Pfleeger and Pfleeger [46], Stamp [59], and Stallings and Brown [58], among others.

2.1   A security taxonomy

This section focuses on describing an information and computer security taxonomy that includes the main types of security properties. Although no agreement exists concerning what would be the reference taxonomy for information and computer security, we rely on a set of security aspects that are most commonly included in security taxonomies, and that existing standards about security do acknowledge.

It is worth mentioning that three of these aspects are widely accepted as essential; these constitute the so-called CIA triad (or security triad): confidentiality, integrity, and availability. The existing debate concerns which other aspects should be added to the CIA triad.

2.1.1   Confidentiality

There is certain information that people and organizations do not wish to disclose, or that they would like to share only with selected people. In our healthcare running example, confidential information includes blood test results, the names of hospitalized people, the identity of donors, the names and addresses of current patients, payment details, and so on.

An important aspect of confidentiality is *privacy*: the capability of individuals to control or influence the disclosure of information that concerns them. For a patient, privacy could concern the capability of deciding whether, with whom, and for what purpose to disclose hospitalization status. Different patients may have rather different privacy requirements; while a hospitalized celebrity may want to disclose no information whatsoever, a hospitalized researcher may be eager to disclose her data to further the research in the field.

In this book, we adapt the definition by Stallings and Brown as follows:

DEFINITION 2.1   Confidentiality ensures that private or confidential information is not made available or disclosed to unauthorized users, and that users control (or influence) what information related to them may be collected and used, and to whom it is disclosed [58].

Note that individual or organizational policies concerning confidentiality may be specified with very different levels of detail. Consider the following examples:

1. Personal data concerning donors are confidential.
2. The names and phone numbers of donors are confidential.
3. The names and phone numbers of donors can be accessed only by doctors.
4. The names and phone numbers of donors can be accessed only by doctors in order for them to investigate a donor's adequacy.

These examples show an increasing level of detail. The first item is unclear on what personal data consist of, and what confidentiality means. The second item clarifies that personal data consist of name and phone number. The third item explains that confidentiality means the ability of doctors to access such data. The last item also specifies that doctors have access only when they need to investigate the adequacy of a donor.

Although information is the key object that confidentiality policies constrain (and the focus of this book), physical objects and areas can be confidential too. This is the case, for instance, with military equipment and rooms that can be accessed only by certain people.

2.1.2 Integrity

Ensuring information confidentiality does not guarantee that the data are not modified in undesired ways. Integrity can be compromised in different ways: cyber attacks to the databases that store the said information, malicious behavior by users having legitimate access, random faults of the database, accidental mistakes that people make when updating the database, communication errors when the information is exchanged between systems, and so forth.

In our healthcare scenario, integrity is key; think of what would happen if the adequacy status of a potential donor was changed from "inadequate" to "adequate," because of a nurse's mistake, a technical fault when the data are transferred from one hospital to another, or a cyber attacker that wants to compromise the reputation of the clinic.

For integrity, we employ Stallings and Brown's definition:

DEFINITION 2.2 Integrity ensures that information is not changed (modified) or destroyed in an unauthorized way [58].

As pointed out by Gollmann [22], integrity is a prerequisite for other security aspects. When attackers attempt to modify a server's file system to gain access to certain data, they are hitting at the *integrity* of that file system in order to circumvent the *confidentiality* policies for the data. Thus, ensuring integrity is often a preliminary activity to guaranteeing other security aspects.

2.1.3 Availability

A key aspect of security is availability, the possibility of using a system, invoking a service, or manipulating some data by authorized users, whenever they need to. Consider the healthcare setting: what would happen if a surgeon who needed to check the medical record of an emergency patient to be operated on was not able to access it, due to the unavailability of the data retrieval service?

We adapt the definition of availability by Stallings and Brown [58]:

DEFINITION 2.3 Availability ensures that a system works promptly, service is not denied to authorized users, and access to and use of information is timely and reliable.

A well-known term that is tightly connected with availability is denial of service, defined by the NIST *Computer Security Incident Handling Guide* [9] as "an action that prevents or impairs the authorized use of networks, systems, or applications by exhausting resources such as central processing units, memory, bandwidth, and disk space." For example, the inability of the surgeon to check a medical record may be due to a cyber criminal attacking the database that contains such data.

2.1.4 Authenticity

One of the most common examples in computer security (in secure communication, more specifically) depicts two actors, called Alice and Bob, who are communicating by exchanging messages. Another actor, called Trudy, attempts to intercept their messages to deliver a message different from the authentic one, by changing the content of the message. In the healthcare scenario, it is necessary for the surgeon to be certain that the records of potential donors are authentic.

Authenticity is the property that holds when these scenarios are prevented from occurring.

We combine the definition of authenticity from the NIST glossary with that by Stallings and Brown:

DEFINITION 2.4   Authenticity is the property of being genuine and able to be verified and trusted [32]. Authenticity is ensured through authentication processes that verify whether users are who they say they are (entity authenticity [58]).

This definition refers to the key mechanism (see Section 2.3) that supports authenticity: the authentication of users to demonstrate that they are who they claim to be, and not someone else. This is crucial, for example, to be certain that a message Bob receives from Alice has actually been sent by Alice, and not by Trudy.

2.1.5   Reliability

Even when a system is available (in other words, reachable), it may be unreliable in the function that it performs. An example of unreliability is when the information system that provides patients' clinical records is being queried by a surgeon for the current donor, and a result is returned that does not correspond to the donor's record.

For reliability, we adopt the characterization by Gollmann [22], which refers to dependability as per the IFIP Working Group 10.4 [33]:

DEFINITION 2.5   Reliability is the property of a system such that reliance can be justifiably placed on the service it delivers.

The definition emphasizes the fact that reliance is *justifiably placed*, thereby calling for the development of security mechanisms (see Section 2.3) that provide certain guarantees concerning the system. In our example, such mechanisms would support the surgeon to gain trust in the correspondence between the returned record and the patient at hand.

2.1.6   Accountability

A major concern in security is the ability to associate occurred actions with the people and systems that are responsible for their execution. Suppose the surgeon is checking the record of a donor, and the information system she is using mentions that the eligibility of a potential donor has been recently modified to ineligible. Accountability (here, knowing who changed the flag value) is a prerequisite for the doctor to understand whether this change has been made by an authorized person.

We take the definition of accountability from the NIST glossary of security terms [32]:

DEFINITION 2.6   Accountability refers to the requirements for actions of an entity to be traced uniquely to that entity (e.g., non-repudiation of a communication that took place).

Examples of the necessity of accountability are before our eyes. Think of email *spoofing*: someone sending an email pretending to be someone else. If full accountability were properly integrated in our email systems, it would be impossible for the sender to pretend to be someone else and deceive the recipient.

2.2   Managing security: threat and risk analysis

Understanding key aspects of security (Section 2.1) does not suffice to ensure security within an organization, or when designing an information system. Security management is a vital activity that has to be continuously conducted to identify existing and future security concerns, to detect weaknesses that can be exploited to violate security, and to define and enact a plan to prevent or react to security issues.

Threat and risk analysis is the process that is concerned with the systematic study of these concerns. To effectively conduct this process, a basic understanding of some key terms is required. We summarize the main elements in Figure 2.1, which are described later in the following sub-sections while explaining the main steps in the threat and risk analysis process:

1. The assets in the context under study are identified as well as the threats that affect those assets (Section 2.2.1).
2. The vulnerabilities that can be exploited to harm the assets are evaluated (Section 2.2.2).
3. The risk value is assessed, based on the results from the previous steps (Section 2.2.3).
4. Countermeasures are identified to reduce risk value (Section 2.2.4).

Despite the sequential presentation, threat and risk analysis is an iterative process that requires continuous reevaluation. Indeed, new vulnerabilities are typically discovered, the relevant assets may change, new countermeasures are proposed, existing countermeasures may prove to be ineffective, and so on.

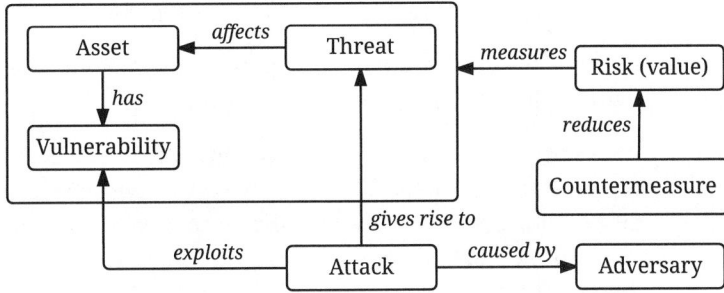

Figure 2.1
Main concepts in the threat and risk analysis process.

2.2.1   Identification of assets and threats

Threat and risk analysis starts with identifying the entities that the organization (wherein the designed information system will operate) values, and that should therefore be protected. These valuable entities are known as *assets* and are defined as follows:

DEFINITION 2.7   An **asset** is anything that has value to the organization and that therefore requires protection [29].

This definition is very general and does not specify exactly the types of entities that have to be protected. In the context of this book, which takes a socio-technical perspective on security, we consider *technical* assets such as software and computers, *information* that organizations rely on to function properly, *processes* that define how organizations deliver their functions, and *people* involved in the conduction of those processes.

The healthcare example contains examples of each asset type: the medical information system that doctors and nurses consult to retrieve patient data (technical), the suitability of a donor and her identity (information), the process of conducting a transfusion (process), and the surgeon, nurse, and donor involved in these processes (people).

Assets are relevant for the security realm because, in real life, many events exist that can break the security of these assets: their confidentiality, integrity, availability, and so on. Such events may occur deliberately or accidentally. Threats can be defined as follows:

DEFINITION 2.8   A **threat** is a potential cause of an incident that may result in harm of systems and organizations [29].

Examples of threats are the disclosure of data (e.g., the donor's identity is revealed to the donee), service interruptions (e.g., the medical information system cannot be accessed to retrieve patient records), unauthorized modification of data (e.g., the eligibility of a potential donor is changed by a secretary who has no rights to do so), loss of data (e.g., the clinical record of a patient is lost).

Two key outputs of this step are a list of assets and a list of threats that affect those assets. The risk analyst typically employs existing catalogs of assets and threats as a starting point [7, 29, 65]. Moreover, the analyst associates the threats with the assets to denote which threats affect which assets. Such association also includes an estimation of the threat severity; for example, a qualitative low-medium-high scale can be employed.[1] The unavailability threat applied to the medical information system asset would be valued as high because of the severe repercussions of such a disruption.

2.2.2   Finding and assessing vulnerabilities

This step is concerned with identifying and evaluating concrete vulnerabilities that can be exploited in order to realize a threat. In other words, it is the exploitation of vulnerabilities that enables the occurrence of a threat. If no vulnerabilities exist, threats will not come into existence.

For example, if the medical information system is fully available, the threat of unavailability cannot be realized. On the other hand, a vulnerability exists if any secretary is allowed to change the eligibility flag of potential donors without doctor approval. Thus, a vulnerability is defined as follows:

---

1 Richer qualitative scales with more points or quantitative values can be employed too, without affecting the overall process we describe.

DEFINITION 2.9   A **vulnerability** is a weakness in an information system, system security procedures, internal controls, or implementation that could be exploited [52].

Definition 2.9 encompasses not only technical/implementation flaws in information systems, but also weaknesses in the organizational policies (*internal controls*), such as the possibility for every employee to access any room, even some where confidential documents are stored.

The notion of vulnerability prompts for two additional concepts describing the intentional attempt to exploit a vulnerability: *attack* and *adversary*. These concepts help the analyst focus on *who* has an interest in exploiting a vulnerability, and *how* these subjects may plan to do so. For example, a possible adversary in the healthcare domain could be a group of activists that oppose blood transfusion, and their attack may be a denial of service against the medical information system.

DEFINITION 2.10   An **attack** is any kind of malicious activity that attempts to collect, disrupt, deny, degrade, or destroy information system resources or the information itself [52].

DEFINITION 2.11   An **adversary** is any individual, group, organization, or government that conducts or has the intent to conduct detrimental activities [5].

Each of the identified vulnerabilities is then analyzed by assessing (i) the likelihood of occurrence, and (ii) the likelihood of success. Analysts typically employ qualitative (low-medium-high) frameworks. For example, the denial of service attack above could have medium likelihood of occurrence due to a growing trend against transfusions, and low likelihood of success because a scalable cloud architecture is used.

These likelihoods are finally combined into a single value through the use of a matrix such as the one shown in Table 2.1. In our example, the vulnerability would be assessed as *low*. Should the medical information system have no scalable architecture, the likelihood of success could be medium, and the vulnerability would be assessed as *medium*.

Table 2.1

A possible matrix to combine vulnerability exploitation likelihoods of attempt and success.

	Likelihood of attempt		
Likelihood of success ↓	Low	Medium	High
Low	Low	Low	Medium
Medium	Low	Medium	High
High	Low	Medium	High

2.2.3   Risk assessment

All the information collected in the previous two steps is now combined into a unified metric called *risk*, or *risk value*, which measures the quantified effect of a threat being realized. Specifically, this puts together the evaluation of the threat-to-asset impact from Section 2.2.1 with the vulnerability value from Section 2.2.2.

DEFINITION 2.12   **Risk** is a measure of the extent to which an entity is threatened by a potential circumstance or event and is typically a function of (i) the adverse impact of the circumstance or event occurring and (ii) the likelihood of occurrence [52].

Risk assessment is often performed in a similar way to vulnerability assessment, by using a matrix such as that of Table 2.2, to determine risk value for every pair (threat-to-asset, vulnerability).

For the example concerning the unavailability (threat) of the medical information system (asset) by the conduction of a denial of service (attack) by the activists opposed to blood transfusions (adversary), the matrix would combine impact *high* (see Section 2.2.1) with vulnerability *low* (see Section 2.2.2), thereby leading to a *medium* risk value according to Table 2.2. If the vulnerability level were *medium*, the overall risk value would be *high*.

2.2.4   Risk response: countermeasures

Risk assessment provides the analyst with a prioritized list of risks on the basis of risk value. In our examples, risks labeled *high* are those with the highest priority, while those labeled with *low* are the least severe for the organization.

Table 2.2

A possible matrix to assess risk value.

Vulnerability ↓	Threat-asset impact		
	Low	**Medium**	**High**
Low	Low	Low	Medium
Medium	Low	Medium	High
High	Low	Medium	High

The following step consists of identifying countermeasures for the identified risks, that is, security mechanisms that can be deployed to tackle the risk. The *National Information Assurance Glossary* provides a comprehensive definition of countermeasures:

DEFINITION 2.13 **Countermeasures** are actions, devices, procedures, or techniques that meet or oppose (i.e., counter) a threat, a vulnerability, or an attack by eliminating or preventing it, by minimizing the harm it can cause, or by discovering and reporting it so that corrective action can be taken [52].

Some countermeasures are organizational and change the security policy or procedures: for example, requiring the secretary to consult a doctor before promoting a potential donor to eligible. Other countermeasures are technical and comprise the installation of devices or the implementation of software/algorithms: for example, installing a firewall to protect the medical information system from denial of service attacks, or implementing an access control module for the medical information system to prevent secretaries from promoting a potential donor to eligible.

There are various ways of tackling risk, which can be combined to increase the efficacy of risk reduction:

- *Eliminate* by removing the activity that gives rise to the risk. This is often an impractical strategy, especially when the core organizational processes or systems are involved.

- *Transfer* the responsibility to another subject. This is typically done by establishing an insurance policy.

- *Accept* the risk and do nothing. This is a viable solution if countermeasures are too expensive, the risk impact is low, or the vulnerability has low likelihood of occurring or being successfully exploited.

- *Prevent* via mechanisms that reduce the likelihood of occurrence.
- *Detect* the risk through the use of monitoring mechanisms deployed in the organization.
- *React* to the risk after its occurrence, by mitigating its impact and recovering from the effects the attack has produced.

For the unavailability risk of the medical information system due to denial of service, a plausible risk response would involve prevention mechanisms such as deploying the system on an elastic cloud platform with adaptive capacity, detection through the use of a firewall that monitors the traffic looking for patterns, and reacting to the occurrence by rerouting the legitimate traffic to a different machine.

The choice of the specific countermeasures to employ depends not only on their efficacy, but also on their cost effectiveness; many countermeasures, indeed, require the acquisition or development of expensive infrastructures that may be too costly for risks with low impact or chance of occurrence.

2.3   Security mechanisms

In the previous sections, we informally mentioned some security mechanisms that can be employed to counter risks and to establish computer and information security. The purpose of this section is to provide a concise overview of the key security mechanisms. Because the focus of the book is on *requirements*, and not on *solutions*, however, this section cannot be but limited. The interested reader can refer to the extensive treatment of these mechanisms found in books concerning computer and information security [22, 46, 58, 59].

A major category of security mechanisms concerns the confidential exchange of messages among two communicating actors. The mechanisms that support this security aspect rely on **cryptographic algorithms**: the sender Alice encrypts her message so that its content cannot be understood by anyone but the recipient, transmits it to Bob through some communication channel, and Bob decrypts the received message to retrieve the original message. The encryption and decryption functions have at least two parameters: the actual message to encrypt/decrypt, and an additional string called *key*.

Up to the late 1970s, all algorithms were based on **symmetric encryption**: the exchanged messages were encrypted and decrypted using the same shared key that both Alice and Bob would possess. The main drawback of symmetric encryption is that anyone possessing the shared key can (i) encrypt a message, pretending to be Alice, and (ii) decrypt a message, reading something only Bob would be allowed to read. This is not a minor problem, considering that Alice and Bob need to share the key in the first place; unless they are physically co-located, how can they do that in a confidential way? With symmetric encryption, they would require another key to encrypt and decrypt the first key.

Three well-known symmetric encryption algorithms, which are still in use in certain contexts, are the Data Encryption Standard (DES), triple DES, and the Advanced Encryption Standard (AES), having increasing robustness. Their robustness is defined with respect to attackers trying to guess the key through brute-force attacks: since they have standard encryption and decryption algorithms, if Trudy intercepts a message, she can attempt to decipher it by trying all possible keys until the decryption leads to a meaningful text. The speed of modern computers has made DES (and triple DES) too weak to protect the message, while AES is strong enough for modern computers (including any foreseeable advance in the years to come).

Public key encryption is a complementary family of techniques that relies on advanced mathematical functions, and its key property is asymmetry: there is no shared key for both sender and receiver. The shared key is replaced by a couple of keys; every person possesses a *private key*, which is kept secret, and a *public key*, which is made available publicly. These keys are complementary: a message encrypted with the private key can be decrypted with the public key, and vice versa.

Public key encryption is useful to provide confidentiality. Consider Alice sending a message to Bob:

1. The content of the message is encrypted by Alice using Bob's public key, which is publicly available.
2. The ciphered version of the message is transmitted to Bob through a communication channel.
3. When Bob receives the message, he uses his private key to decrypt it and obtain the original message. Note that confidentiality is guaranteed because Bob is the only one who possesses the private key.

Another important property to guarantee is **message authentication**. An exchanged message is authentic when its content is genuine (it has not been modified by other parties) and it has actually been transmitted from the declared sender. Thus, message authentication is a protection against active attacks that aim to deliver messages with corrupted contents, with the attacker pretending to be someone else. *Data authentication* is a similar notion that focuses on the authentication of stored data rather than exchanged.

Symmetric key encryption is inadequate for message authentication. An attacker may indeed compromise authenticity by shuffling the message content; upon decryption, the receiver would obtain a message that does not correspond to the one the sender sent.

Alternative techniques exist to support message authentication; the main types of algorithms are the following:

- **Message Authentication Code (MAC)**: a shared secret key is used to produce a small block of data that is appended to the message. The MAC is computed by a function that takes as input the message to transmit and the shared key. The receiver applies the same function to the message and compares the appended MAC with the one the receiver has computed.

- **One-way hash function**: this type of function takes a message in input, adds a padding to increase the message size to a default length, and computes a hash value. Hash functions should be non-invertible; it should be computationally impossible to reconstruct the actual message from the hash. The hash is appended to the message as in the MAC case, and the receiver applies the same hash function and compares the result with the received hash. To provide authenticity, symmetric or asymmetric encryption techniques are applied to the hash by both sender and receiver.

- **Public key encryption**: the message is encrypted using the sender's private key; the receiver uses the public key of the sender to decrypt the actual message. This algorithm also supports **data integrity**, for only the sender would be able to modify the message and encrypt it with her private key.

Note that these algorithms alone provide authentication but do not support confidentiality. It is possible to combine them with further encryption techniques so that confidentiality is supported as well. However, authentication without confidentiality is often used in practice: when exchanging messages, their outputs constitute a **digital signature** of the sender.

Public key encryption requires the development of functions to ensure **key management**. The fact that public keys are publicly available creates the danger of a malicious user forging a key, pretending the key is someone else's. To address this issue, the typical solution is to create a *certificate* issued by a third party that is trusted by the community, typically an institution.

A widely employed mechanism to ensure the confidentiality and integrity of data or systems is **access control**. The purpose of this technique is to regulate the kind of *access actions* that an active entity (called *subject*) can perform on a passive entity (*object*). For example, access control is used to determine whether a nurse trying to retrieve a donor's data from an information system can access the data. The most fundamental access actions are to observe (read) the object and to alter it. More specialized access actions exist, as will be shown in Chapter 3. Access control is typically enacted by a software infrastructure that processes the access actions and returns whether they can be executed.

User authentication concerns the provision of adequate credentials to convince another party that a user is who she claims to be. There are different means to prove an individual's identity [58]:

- Something the individual knows, like a password she keeps private, a personal identification number such as those used with payment cards, or answers to predefined questions.

- Something the individual possesses, like a smart card or a physical key.

- Something the individual is, including the recognition of biometric parameters such as fingerprint, retina, and face.

- Something the individual does, such as recognition by voice pattern or handwriting.

Availability can be maximized through the use of **firewalls**, which are devices (hardware or software) that stand between a system within a given network and users or systems outside the network. Firewalls act as a filter for the incoming and outgoing traffic: they decide whether requests to the system and messages coming from the system should be allowed. These decisions rely on various parameters: the Internet address of the entity outside the network, the communication port that is employed, the content of the message, and so forth.

Reliability is boosted by means of *redundancy*, a notion that is adopted from reliability engineering. Redundancy consists of duplicating critical components or functions of a system to increase its reliability (i.e., to lower the likelihood of system failure). A common example of redundancy concerns hard disks: the RAID (Redundant Arrays of Independent Disks) configuration replicates the same data over multiple disks, so that the failure of one of them does not lead to the loss of data.

General mechanisms contribute to supporting multiple security aspects:

- Technical mechanisms: *intrusion detection systems* constitute a holistic tool to ensure security in an organization; they sense data, perform an analysis to detect if a possible threat exists, and trigger reactions by coordinating other security systems within the organization. *Antivirus* and *antispyware* play a similar role on individual computers, by monitoring files and user activity, and enacting reactions when required.

- Organizational mechanisms: *security maturity models* are frameworks that analysts employ to assess the maturity level of an organization concerning security, and to determine areas for improvement. The Software Assurance Maturity Model [45] provides recommendations for an organization to develop secure software systems by looking at various facets of security for software systems: governance, construction, verification, and deployment. The International Standards Organization and other standardization bodies have proposed several frameworks to examine security within organizations, such as the ISO/IEC 27005:2011 standard on information security risk management [29], the information security part of ISACA's framework for governance and management of enterprise IT [27], and so forth. These mechanisms are often instantiated by external independent analysts that perform *security audits* by studying the organization in terms of its security policies and systems.

2.4 Chapter summary

This chapter has provided an overview of computer and information security that establishes some key terminology used in this book and constitutes a brief outline of the field. The chapter is not intended, however, to replace existing textbooks on computer and information security. The reader should have learned a few basic notions about security:

- Many complementary security aspects exist. In this book, we employ a taxonomy that consists of confidentiality, integrity, availability, authenticity, reliability, and accountability.

- The fundamental approach to security involves threat and risk analysis, as explained in Section 2.2.

- Risks are dealt with by developing and enacting security mechanisms. While several of these mechanisms are technological, others are organizational and focus on the adherence of people to security guidelines.

2.5 Exercises

Review questions

Q2.1. Explain what information assets are, and provide some examples in the context of a bank.

Q2.2. What is the difference between confidentiality and privacy?

Q2.3. Consider the security aspect of authenticity. Explain the difference between user authentication and data authentication. What are the mechanisms to ensure authenticity?

Q2.4. What is accountability? Why is it an important security notion?

Q2.5. How do threats and vulnerabilities relate with each other?

Q2.6. Define the notion of risk, and explain how to derive this value.

Q2.7. What is public key encryption? How does it differ from private key encryption?

Q2.8. What are the key elements of access control? What security aspects does access control contribute to?

II THE STS-ml MODELING LANGUAGE

3   The Socio-Technical Security Modeling Language

This chapter presents a modeling language for representing security requirements in *socio-technical* systems, namely the Socio-Technical Security modeling language (STS-ml). The proposed language is the central artifact of the model-driven approach to SRE that this book promotes.

Section 3.1 outlines the ten principles that STS-ml relies on. Section 3.2 describes the modeling primitives that represent the stakeholders in a socio-technical system. Section 3.3 details the primitives for the interactions among those stakeholders. Section 3.4 explains how to model the events that threaten the stakeholders. Finally, Section 3.5 describes the security requirements that STS-ml supports.

3.1   The ten design principles for STS-ml

Based on Chapter 2's background on computer and information security as well as the motivations for a new approach described in Section 1.4, we define a number of design principles for STS-ml. These are intended to ensure the language's fitness for its application domain (security requirements for socio-technical systems) and to guarantee that it accommodates some desired properties (requirements) for modeling languages.

Principle 1 Socio-technical perspective. *The modeling primitives must enable creating security requirements models for a socio-technical system that encompasses humans, organizations, and software systems. These participants interact to fulfill their own strategic objectives (humans and organizations) and requirements (software systems).*

Security requirements for software systems, such as "The system shall ensure the confidentiality of the collected donor's personal data" (R_1), have to be represented by primitives that acknowledge the socio-technical nature of the setting. This means representing the existence of multiple actors (e.g., a donor, a nurse, and the hospital's information system), their interaction for the purpose of donating blood, the need to record the donor's personal data, and her desire to keep such data confidential.

Principle 2 Security about interactions. *The participants in a socio-technical system are autonomous and, thus, loosely controllable.*

While interacting (to get things done and to transfer information), however, a participant often imposes security requirements on the interacting actors.

Let us consider security requirement R_1 again and its socio-technical interpretation as sketched in the previous paragraphs. The desire to keep the transferred data confidential is an example of a security requirement that is imposed by the donor on her interaction with the nurse, and this requirement affects in turn the interaction between the nurse and the information system. Indeed, if the nurse does not mark the data as "confidential," or if the information system does not allow this option, the donor's need could potentially be violated.

Principle 3 High-level assets. *The language should support modeling the high-level assets stakeholders care about, such as their strategic objectives and their information.*

(Security) requirements engineering is conducted at the initial stage of system design. As such, the activities in this phase employ a higher level of abstraction than those in following phases. A key principle in computer and information security is the protection of the assets the stakeholders value. The language should support representing these assets at a high level of abstraction to express security requirements for their protection. In the example of R_1, a donor's asset is her personal data, while a hematologist's asset is the goal of assessing the adequacy of prospective donors.

Principle 4 Threats. *Security analysis often includes the perspective of possible attackers that aim to exploit the vulnerabilities of the system. Socio-technical systems are no exception; therefore, our language should not overlook threats and should support the identification of social and organizational threats, which do not exploit technical vulnerabilities.*

In the example of R_1, possible threats are that the nurse forgets to mark the donor's personal data as confidential, or that she leaves a printout of the donor's record in the waiting room. Notice that these threats are social/organizational, for they do not exploit a technical vulnerability of a software system.

Principle 5 Multiple stakeholders. *By their own nature, socio-technical systems consist of multiple participants. Each participant is* *an autonomous stakeholder and, as such, specifies her own (security) requirements, which may conflict with others'. The language should therefore express the stakeholders' viewpoints and facilitate the identification of conflicts.*

Security requirement R_1, expressed by donors, may conflict with the requirement of hematologists to access some details, such as age, profession, drinking/eating/smoking habits, and so forth. The language should be able to represent these viewpoints as well as identify when a conflict occurs.

Principle 6 Diagrammatic and formal language. *The language has to support modeling through diagrams and provide formal and unambiguous semantics for its primitives.*

While (security) requirements engineering is often conducted through informal analysis and relies on textual requirements descriptions, this book advocates a model-driven approach where models are represented through diagrams, which serve as a communication means among modelers and with stakeholders. These diagrams have an underlying formal semantics (i) to minimize ambiguities in their interpretation and (ii) to enable performing automated reasoning to detect conflicts among requirements.

Principle 7 Compliance with standards. *Whenever possible, the primitives in the language should adhere to standards and common terminology, so as to improve comprehensibility and to avoid a steep learning curve.*

This principle applies to many fields and disciplines. In the context of this book, the challenge is to reuse and adapt, in the context of socio-technical systems, the terminology used in traditional security for software systems. The approach proposed in this book relies on the six aspects summarized in Table 3.1, which are assembled from mainstream taxonomies and security standards as detailed in Section 2.1.

Principle 8 Minimality of concepts. *The language should be composed of a minimal set of primitives—concepts and relationships—that are needed to capture security requirements in socio-technical systems.*

Table 3.1

Taxonomy of computer and information security employed in this book.

Aspect	Definition
Confidentiality	Ensures that private or confidential information is not made available or disclosed to unauthorized users, and that users control (or influence) what information related to them may be collected and used, and to whom it is disclosed [58].
Integrity	Ensures that information is not changed (modified) or destroyed in an unauthorized way [58].
Availability	Ensures that a system works promptly, service is not denied to authorized users, and access to and use of information is timely and reliable [58].
Authenticity	Is the property of being genuine and able to be verified and trusted [32]. Authenticity is ensured through authentication processes that verify whether users are who they say they are (entity authenticity [58]).
Reliability	Is the property of a system such that reliance can be justifiably placed on the service it delivers [22].
Accountability	Refers to the requirements for actions of an entity to be traced uniquely to that entity [32] (e.g., non-repudiation of a communication that took place).

This principle requires that the chosen primitives not be redundant and that they enable representing the important security requirements that the stakeholders wish to express.

Principle 9   Traceability. *The language should ensure traceability of security requirements to their requester and to the motivations/goals that originated the requirement.*

For example, considering R_1, the language should enable determining who requested the requirement (the donor), and why (the donor wants to have her blood examined and needs to provide her personal data to that extent).

Principle 10   Capturing security needs. *The language should focus on the needs of the stakeholders, while designing socio-technical systems with security in mind, rather than on the solutions or mechanisms to address these needs.*

The language is thought for an early stage of system development. As such, its models should express why security is needed and what the security needs of the stakeholders participating in the system are.

Security solutions or mechanisms are better considered in later stages of system design.

3.2 Representing actors in socio-technical systems

Socio-technical systems consist of multiple autonomous interacting participants: humans, organizations, and technical components. Software systems and infrastructures are examples of technical components. Organizational units and employees, on the other hand, are examples of social components. As per Principle 1, the language should enable representing these participants.

STS-ml does so through the modeling primitive of an *actor*, that denotes a participant carrying out actions (a *doer*). Examples of actors are a donor, a doctor, a nurse, the authors of this book, and its readers.

Actors in STS-ml are used to represent the notion of *stakeholder*, which is a central one in requirements engineering. Stakeholders are indeed the subjects that express their needs on the system-to-be.

In this book, the terms stakeholder, participant, and actor are used interchangeably, as they refer to different facets of the same entity. Specifically, the modeling language represents the socio-technical system in terms of actors, which correspond to the stakeholders that express their needs about the system-to-be on behalf of the actual participants in the system.

For example, a sample set of donors would be selected and constitute the stakeholder "Donor," which expresses requirements on behalf of the prospective participating donors. A donor would be represented as an actor in STS-ml.

3.2.1 Actor types

The notion of actor is a very generic one, for many different entities can be classified equally as actors. For instance, "Physician," "Physician Mark," and "Mark" can all be considered actors. While a complete treatment of this subject is outside the scope of this book (for a comprehensive and well-founded account on the topic, see [38] and [23]), STS-ml includes two concrete refinements of actor that modelers can use: role and agent.

A role is an abstract characterization of an actor, and it is used to model actor responsibilities. One can think of role as a container that carries a set of responsibilities that the actor has within the system. Roles are used to model participants/actors when the actual individual(s) are unknown. Examples of roles include Patient, Physician, Donor, and so on. One knows that patients, physicians, and donors will be part of the system, can define their responsibilities, but does not necessarily know the identity of the agents that will play those roles.

An agent, on the other hand, refers to a specific individual that will participate in the system. For example, St. John's Hospital and Hong Kong Red Cross are agents that will participate in the blood donation socio-technical system. Roles and agents in STS-ml are graphically represented in Figure 3.1.

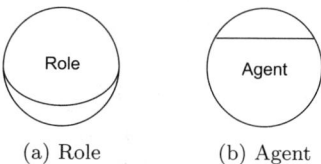

(a) Role   (b) Agent

Figure 3.1
Graphical representation of role and agent.

Agents and roles can be related one to another: an agent can play (adopt) a role. For example, Mark can play role Physician, while Alice may play role Donor. STS-ml introduces the relationship plays between an agent and a role, graphically shown in Figure 3.2.

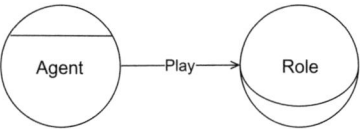

Figure 3.2
Graphical representation of a plays relationship.

Some observations about the plays relationship are worth noting. An agent can play multiple roles, and multiple roles can be played simultaneously. For example, one of the authors of this book is both an advisor and a coauthor for another author. In other cases, an agent changes its role in the socio-technical system. For instance, Alice may play role Nurse and, after work, switch to role Donor.

3.2.2 Actor assets

A fundamental element of security is the identification and protection of the assets stakeholders care about. As pointed out in Principle 3, STS-ml is concerned with high-level assets that can be identified at requirements time, before design decisions about the system-to-be are made.

STS-ml takes the standpoint that every stakeholder (thus, every modeled actor) comes with a set of assets that it wants to protect, its *primary assets* [29]. These assets are threatened when they relate to tangible/-concrete assets that are vulnerable, which are known as *supporting assets* [29]. For instance, a Donor's personal information and health status are her primary assets, while the donor's certificate is a supporting asset that contains this relevant and possibly confidential information.

Whenever a supporting asset is handled in an unauthorized way, the primary assets it is related to could be harmed too. Therefore, the necessity of protecting supporting assets depends on the criticality of their corresponding primary assets. If the certificate did not contain any health-related data or personal information, the Donor would not consider its protection to be critical.

STS-ml is concerned with the security issues that actors face in the socio-technical system, and therefore two main types of assets are considered: *informational* and *intentional*. This categorization is related to the fact that actors in a socio-technical system enter with the intention of achieving their objectives, and they interact with others while exchanging related information.

3.2.2.1 Informational assets

Actors are concerned with the information they own, which they often consider confidential (it is a primary asset for them). STS-ml considers actors' information to be made available in the form of documents. A document represents a transferable entity (e.g., donor certificate, identity document, driver's license, an email), which *may* contain some information (e.g., date of birth, name, surname, address, medical status). Since documents represent information, they are supporting assets. Hence, to protect information from misuse, it is crucial to protect the documents that make this information available (materialize) and enable its transmission [29].

Understanding the distinction between information and document is crucial for a proper use of STS-ml. The question that a modeler should pose is "Can I transfer entity X as is?" If the answer is affirmative, then entity X is a document, and most likely the document will represent some information. If the answer is negative, then X is information, which needs to be represented through some transferable entity before being transferred. The two concepts are illustrated in Figure 3.3.

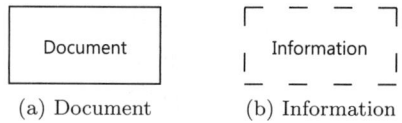

(a) Document   (b) Information

Figure 3.3
Graphical representation of informational assets.

Consider some examples to better understand the difference between information and document. An individual's driver's license number is an instance of information, which is transferred when concretely represented through a driver's license card (think of when one exhibits this card to a police officer). Similarly, nationality is the legal relationship between a person and a state; as such, it is an example of information. One's nationality is transferred through documents such as the passport that one shows to border agents, or an electronic certificate or social security card one sends to a prospective employer. A third example concerns credit cards; the number itself is an instance of information, which can be transferred by swiping the credit card, by typing the number into an online form by sending an email to a merchant.

Documents are possessed by actors, and STS-ml represents this through the possesses relationship, which denotes that the actor may dispose of a document in the socio-technical system. For instance, Red Cross BTC possesses donor certificates. The graphical representation of actor possession is shown in Figure 3.4a. Actors can manipulate only the documents they possess. Moreover, they can get possession of a document when transferred from some other actor. Document manipulation and transmission are explained in Section 3.2.3 and Section 3.3, respectively.

Information has one or more legitimate owners. For example, a Patient's personal data are owned by the Patient, a credit card number is owned by the cardholder, and a bank account number is owned by its holder(s). In STS-ml, the relation owns indicates one actor (a role or an agent) is the legitimate owner of some information and can freely dispose of it, as well as decide to transfer rights related to it to others. The graphical representation of information ownership is shown in Figure 3.4b.

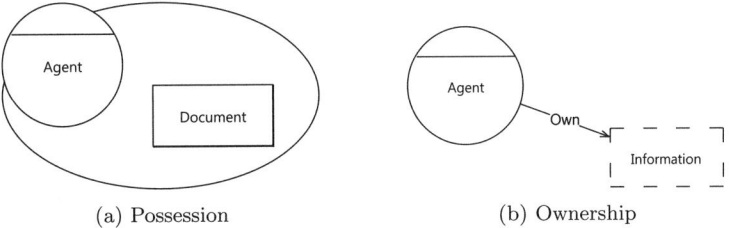

(a) Possession   (b) Ownership

Figure 3.4
Graphical representation of document possession and information ownership.

In order to talk about confidentiality and other security requirements concerning information, it is crucial to model the relationships between documents and information, which are the supporting and primary assets in the category of informational assets. Section 3.2.4 elaborates on how STS-ml enables a fine-grained modeling of these relationships.

3.2.2.2 Intentional assets

The behavior of an agent is a result of its attempt (intention) to fulfill its objectives. Agents want their intentions not to be threatened by other agents with different intentions (including attackers). A role characterizes a specific class of agents; intentions are ascribed to roles too, meaning that any agent playing that role would inherit those intentions.

Note that only rational agents are considered here, that is, agents whose actions are directed toward the achievement of their goals. Technical components are agents whose requirements are expressed via models from goal-oriented requirements engineering [61].

STS-ml represents intentional assets via the notion of goal, which denotes a desired state of affairs (e.g., donor approved, blood donated) for an agent or a class of agents (a role). Actors enter the socio-technical system with the aim to achieve their desired goals; thus, their main goals are their primary assets.

Goals are different from activities and processes: goals are part of the motivational component of an agent; they express *why* and *what* an agent aims to achieve, rather than *how* the agent is to achieve its objectives. Figure 3.5a shows how a goal is graphically represented, while Figure 3.5b illustrates that an actor (in the example, an agent) has the intention of achieving a goal.

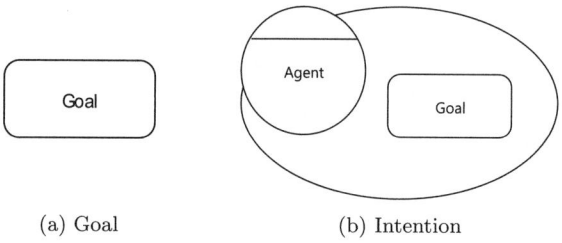

(a) Goal   (b) Intention

Figure 3.5
Graphical representation of goal and intention.

Figure 3.6 shows some intentions for the actors in the healthcare scenario. The agent Red Cross BTC, for instance, has the goals of distributing, collecting, and examining blood. Moreover, it has the goal of performing statistical analysis. The Patient has the goal of receiving a treatment, while the Donor has the goal of performing blood donations on a regular basis. Notice that Alice is an agent who plays role Donor, but her goals are not the same as the role she adopts. In other words, agents may have personal goals that differ from those of the roles they play.

In Figure 3.6, goal labels (when not abbreviated) are in passive voice: treatment received, tests taken, blood distributed, etc. This form is chosen since a goal refers to a desired state of affairs that the actor wants to achieve, as opposed to describing a concrete action to be performed. This shows the flexibility of goal orientation, which focuses on attaining the targets of the stakeholders, rather than on following a detailed procedure.

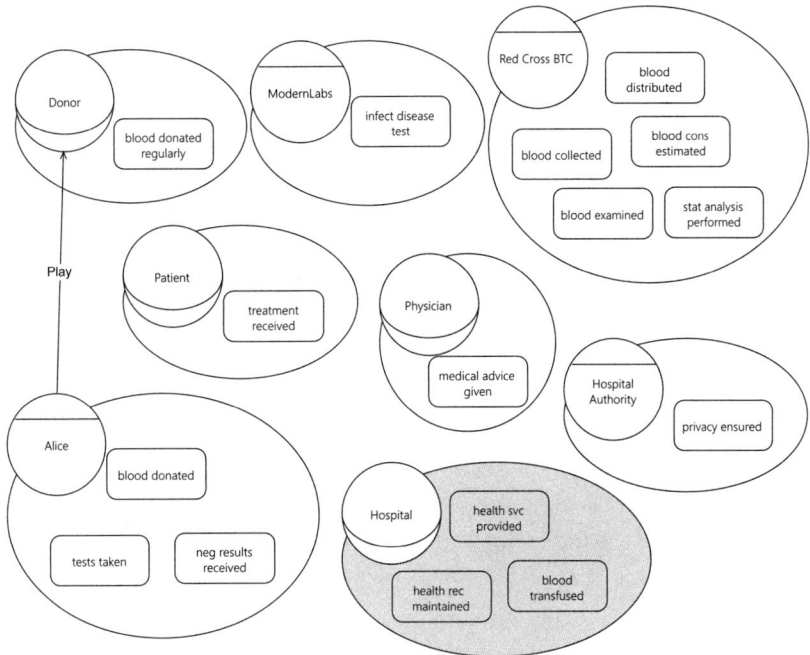

Figure 3.6
Actors' intentions in the healthcare scenario.

3.2.3   Actor models

The goals and documents of a specific actor in a socio-technical system
relate one to another. Important relationships need to be captured that
enable defining the rationale of that actor, that is, how it aims to attain
its goals. The purpose of this section is to explain how STS-ml supports
modeling an actor's rationale.

The rationale of an actor is expressed by the elements and relationships are contained within an *actor model*. This model consists of the
actor's intended goals, its possessed documents, and the relationships
between these elements, which are described in this section. Graphically, an actor model consists of all elements and relationships within
the boundary of an ellipse attached to the actor shape, as in Figure
3.4a, Figure 3.5b, and Figure 3.6.

3.2.3.1   Goal relationships

A fundamental way to relate the goals within an actor model is to refine (*decompose*) them into subgoals. Goal decomposition is widely adopted in requirements engineering [6, 63, 66], and it allows reading goal hierarchies both top-down (to answer the question "*How* is the goal achieved?") and bottom-up (to answer the question "*Why* does this goal exist?").

Inspired by the literature in requirements engineering, STS-ml supports two types of goal decompositions, wherein one goal G is hierarchically refined into a non-empty set of subgoals $\{G_1, \ldots, G_n\}$:

- and-decomposition (illustrated in Figure 3.7a): the achievement of all the subgoals implies the achievement of the decomposed goal G. For instance, the actor model of **Red Cross BTC** in Figure 3.9 includes an and-decomposition of goal **blood distributed** into subgoals **blood collected** and **blood consumption estimated**, meaning that both collection and consumption estimation are necessary for blood distribution.

- or-decomposition (illustrated in Figure 3.7b): the achievement of at least one subgoal implies the achievement of the parent goal. For example, in Figure 3.9, goal **statistical analysis performed** is or-decomposed into subgoals **on blood type eval**, **on hospital requests**, and **on donors**, referring to the different types of statistical analysis that could be performed by the **Red Cross BTC**.

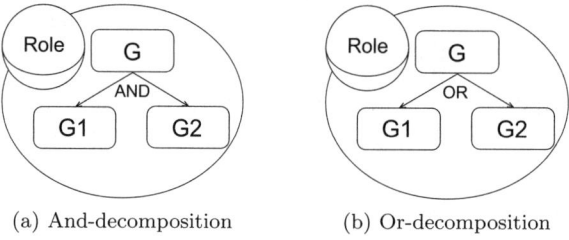

(a) And-decomposition   (b) Or-decomposition

Figure 3.7

Graphical representation of goal and- / or-decompositions.

As shown in Figure 3.9, subgoals can be further decomposed. It is required that a goal have at most one parent. Thus, an actor model is a set of goal trees; each tree refines a top-level (root) goal into a set of leaf-level goals. Since the subgoals are a means to achieve actors' main goals, the set of subgoals constitutes the supporting assets for the actors' corresponding top-level goal (primary asset). Thus, the achievement of leaf-level goals is critical to the achievement of actors' top-level goals.

The refinement of goals in subgoals indicates how actors intend to achieve goals. Refinement continues until there are enough details for the actor to evaluate the achievement of the goal. Think of reducing complex problems to simpler ones. The actor needs to achieve leaf goals first, in order to achieve their parent and ancestor goals, for ultimately reaching the root goals.

Leaf-level goals in an actor model denote the responsibilities of that actor. For a role, these responsibilities apply to any agent adopting that role. For an agent, the responsibilities apply to the concrete participant it represents.

3.2.3.2   Relating goals with documents

Actors often use information to achieve their goals. As explained in Section 3.2.4, information is made accessible through the documents that represent it. STS-ml supports several relationships between goals and documents, which indicate how actors manipulate documents (and, potentially, the information therein) while pursuing their goals. These relationships, illustrated in Figure 3.8, are as follows:

- reads: indicates that an actor reads the content of a document while achieving a goal, thereby making the document necessary for the actor to achieve the goal. For example, in Figure 3.9, document test results is read by actor Red Cross BTC to achieve goal donor approved, for the adequacy of the potential donor to be determined.

- modifies: indicates that an actor changes the informational content of a document while achieving a goal. For instance, in Figure 3.9, document blood bank is modified by Red Cross BTC while achieving goal blood cons estimated (estimating the blood consumption in the center).

- produces: denotes that an actor creates a new document while achieving a goal. For instance, in Figure 3.9, document donor certificate is produced by Red Cross BTC while achieving goal donor approved, as a proof that a certain individual has been recognized as a certified donor.

Figure 3.9 shows an actor model for the Red Cross BTC that illustrates the goals of the actor (e.g., blood distributed, blood collected), its possessed

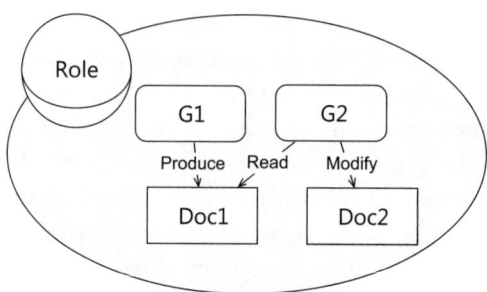

Figure 3.8
Graphical representation of goal-document relationships.

documents (e.g., test results, report), decomposition relationships among the goals, as well as goal-document relationships.

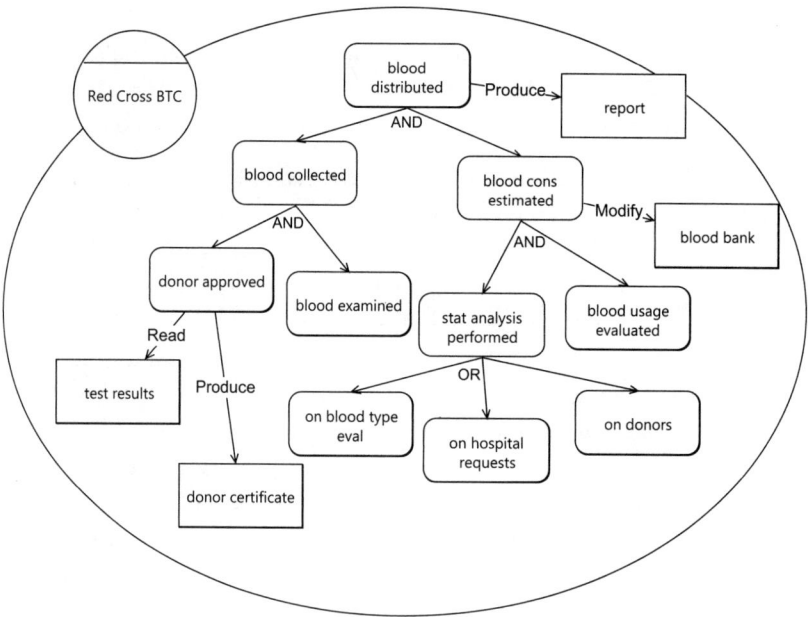

Figure 3.9
Actor model for Red Cross BTC.

As mentioned before, the availability of needed documents is crucial for the achievement of related goals. The set of subgoals and documents necessary for the fulfillment of an actor's top-level goal are the supporting assets for that root goal (primary asset).

3.2.4   Structuring information and documents

Information is an important asset that stakeholders want to protect. It becomes vulnerable when represented in a document (e.g., a database record, an email, a letter, an instant message) that can be accessed and modified by others. STS-ml introduces three primitives (their graphical representation is in Figure 3.10) that allow for structuring information and documents:

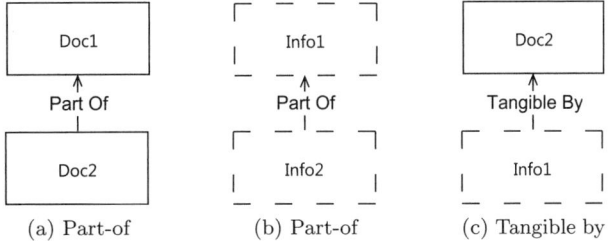

Figure 3.10
Graphical representation of part-of and tangible by.

- part-of between two documents (illustrated in Figure 3.10a) denotes that one document is an essential part of another document (the latter document cannot exist without the former). For example, in Figure 3.11, the report created by the Red Cross while assessing a donor's adequacy is part of the health record of that potential donor.
- part-of between two information entities (illustrated in Figure 3.10b) denotes that certain information is an essential part of other information (the latter information cannot exist without the former). For example, in Figure 3.11, information health status is part of medical history. Also, personal data is part of medical history.
- tangible by between information and a document (illustrated in Figure 3.10c) indicates that the specified information is represented through that document. For example, in Figure 3.11, information medical history

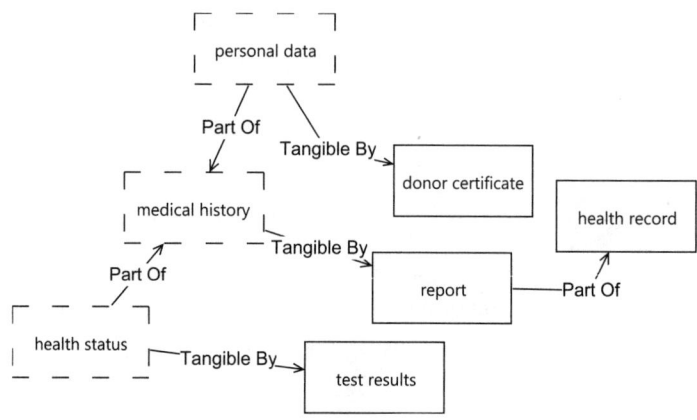

Figure 3.11
Some information-document relationships in the healthcare scenario.

is made tangible by the document report, while health status is made tangible by the test results.

Note that these three relationships are of type many-to-many. For example, an information entity can be part of multiple information entities, and an information entity may have multiple parts. In Figure 3.11, for instance, both health status and personal data are part of medical history.

3.3   Modeling the interactions among actors

The actors in a socio-technical system are not isolated entities, but rather they interact with others either to fulfill objectives they cannot achieve on their own, or to exchange information. Also, an actor may rely on another when it would be capable of achieving a goal, but it finds it more convenient or easier to rely on someone else.

In requirements engineering, the notion of *social dependency* [69] has been suggested as a way to represent that actors rely on others for the achievement of goals, the execution of tasks, and the availability of resources. For example, a Patient may depend on a doctor for goal blood transfused. STS-ml refines the notion of dependency into two specific primitives that are suited for security requirements models:

- goal delegation (illustrated in Figure 3.12a): one actor (delegator) delegates to a different actor (delegatee) the fulfillment of a goal (delegatum). Delegation refines dependency by requiring the existence of an agreement between the delegator and the delegatee, as well as the transfer of responsibility. On the other hand, one actor may depend on another even if the latter has not agreed. In Figure 3.13a, Alice delegates to ModernLabs goal tests taken. By doing so, ModernLabs becomes responsible for testing Alice's blood sample.

- document transmission (illustrated in Figure 3.12b): one actor (sender) transfers to a different actor (receiver) a document. This refines dependency in the sense that the receiver depends on the sender for the availability of the document. In Figure 3.13a, ModernLabs transmits document test results to Alice. In turn, she transmits it to actor Red Cross BTC.

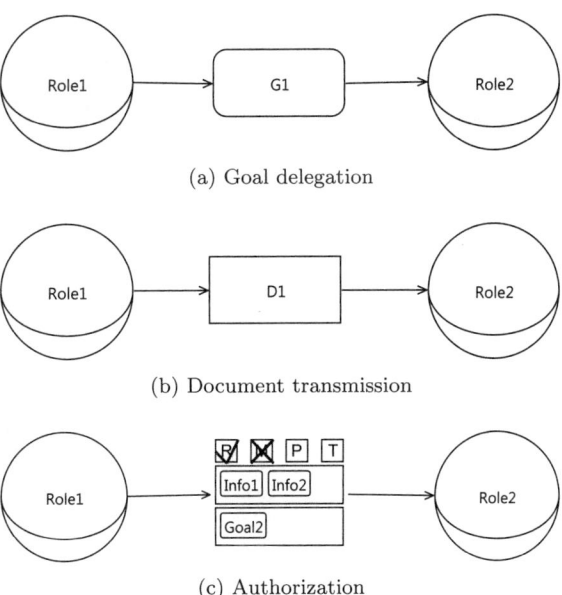

(a) Goal delegation

(b) Document transmission

(c) Authorization

Figure 3.12

Graphical representation of interactions among actors: goal delegation, document transmission, and authorization.

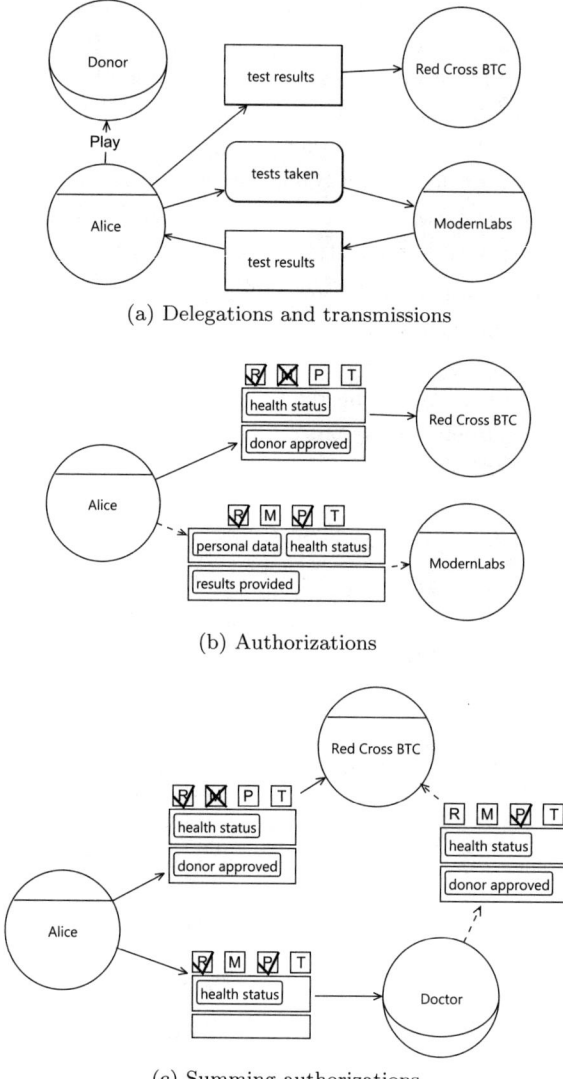

(a) Delegations and transmissions

(b) Authorizations

(c) Summing authorizations

Figure 3.13
Some interactions in the healthcare scenario.

In addition to interacting via delegations and transmissions, STS-ml includes a primitive (authorization, illustrated in Figure 3.12c) to capture two key concepts in security, that is, *permissions* and *prohibitions*. The basic meaning is that one actor (in the example, Role1) specifies authorizations for another actor (Role2) on a set of information elements (Info1 and Info2). For example, in Figure 3.13b, Alice specifies an authorization on information health status for Red Cross BTC, and one authorization on personal data and health status for ModernLabs. Each authorization specifies what the authorized actor can/cannot do with the information:

- *Allowed/prohibited operations*: they define whether the actor is permitted (checkmark symbol) or prohibited (X symbol) to Read (R), Modify (M), Produce (P), and Transmit (T) any document that makes tangible the information. Note that the four supported operations are directly linked to the way documents are manipulated in the social view. A goal can involve reading one or more documents to be fulfilled; an actor can produce (create new) documents during the fulfillment of a goal; an actor might modify a document while fulfilling a goal. A document is modified if, despite the change or update, the document's identity is unvaried. Note also that in STS-ml, operations are complementary; that is, no operation among R, M, P, or T is subsumed by any other operation from the same set. To exemplify this, consider that an actor does not necessarily need to know the contents of the document (read) in order to modify (write, update) it. For instance, a user may add files to a folder (modifying the folder's contents) without necessarily seeing other files contained in that folder. Therefore, in STS-ml the security analyst should specifically and explicitly model which operations are allowed or prohibited.

  In Figure 3.12c, reading is allowed, modification is prohibited, and no authorization is specified on production and transmission. In the healthcare scenario (Figure 3.13b), the authorization from Alice to ModernLabs grants permission to read and produce documents that make tangible her personal data and/or health status; the authorization from Alice to Red Cross BTC permits reading but prohibits modification of documents containing her health status.

- *Scope of authorization*: it defines the goals for the fulfillment of which the authorization is granted. In other words, the authorization is restricted to a certain purpose and does not apply to different purposes. In Figure 3.12c, the permissions and prohibitions are restricted to the scope of goal Goal2. In the healthcare scenario of Figure 3.13b, the agent Red Cross BTC receives authorizations in the scope of goal donor approved; therefore, the permission to read and the prohibition to modify Alice's health status apply only when Red Cross BTC is carrying out activities concerning the approval of Alice as a Donor.

- *Transferability of the permissions*: it specifies whether the actor that receives the authorization is in turn entitled to transfer the received permissions or specify prohibitions (concerning the received permissions) to other actors. Graphically, transferability of the authorization is allowed when the line connecting the two actors is solid, while it is not granted when it is dashed. In Figure 3.13b, Alice allows the transferability of the authorization granted to Red Cross BTC, while she does not allow the transferability of the authorization granted to ModernLabs. Therefore, Red Cross BTC can authorize other actors, specifying both permissions and prohibitions concerning the reading of information health status in the context of goal donor approved.

Note that authorizing some actor over some information means that the actor is authorized for parts of information as well, because ownership of information propagates over part-of relationships. Similarly, when authorization is restricted to a specific goal, it is granted not only for that goal, but also for all of its subgoals, because authorizations propagate over goal trees.

Moreover, authorizations are summed up, that is, whenever different authorization relationships are drawn toward one given actor, then it is enough that at least one authorization grants a given right (say, to read) and none prohibits the same right. For instance, in Figure 3.13c, actor Red Cross BTC obtains two authorizations for information health status: together, they specify that Red Cross BTC can use and produce documents representing health status and is prohibited from modifying any document representing that information.

3.4   Events and threats

Following Principle 4, our language supports the representation of threats in terms of events that exploit the vulnerabilities of actors' supporting assets (subgoals and documents) in order to undermine their primary assets (root goals and information). The primitives of STS-ml are the entity Event and the relationship threatens, which links an event to a document or a goal.

The graphical representation of events threatening actors' assets for the two categories is shown in Figure 3.14. Figure 3.15 provides two examples of events threatening actors' assets, one for each category. The event Physician sick threatens the goal transf needed (Figure 3.15a), as the Physician treating a Patient is the one who evaluates whether a transfusion is needed. The event test results lost threatens the document test results produced by ModernLabs (Figure 3.15b). This event affects ModernLabs in providing a timely service to its patients, in this scenario, to Alice.

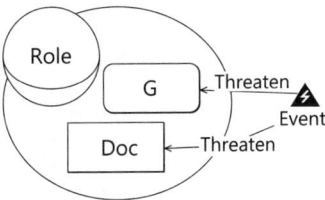

Figure 3.14
Graphical representation of events threatening actors' supporting assets.

A key component of information security is to conduct risk assessment, which enables identifying events and the threatened supporting assets. STS-ml does not propose a specific method in particular but advocates using some method to identify events and threatens relationships. More details are provided in Chapter 6 when describing the STS method.

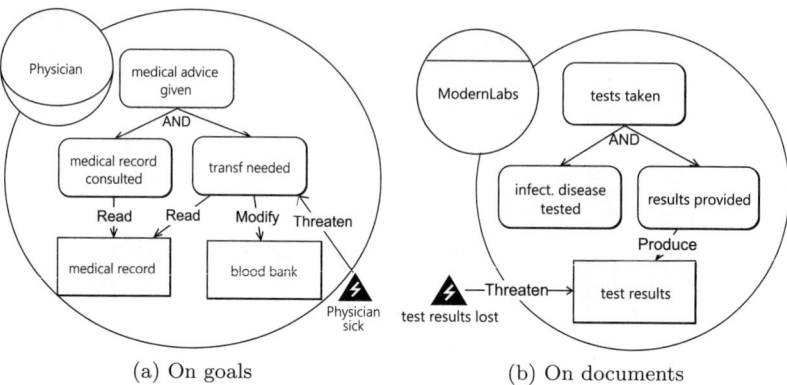

(a) On goals   (b) On documents

Figure 3.15
Some threats affecting actors' supporting assets.

3.5   Expressing security requirements in STS-ml

The main aim of STS-ml is to represent the security requirements expressed by the stakeholders of the socio-technical system (the modeled actors). As shown earlier in this chapter, a preliminary step is to represent the structure of the socio-technical systems in terms of actors, their goals, documents and information, and social interactions.

Guided by Principle 7, the language should express security requirements using a terminology that security experts are already familiar with. Unfortunately, there is no agreement on a reference taxonomy of security requirements and mechanisms. This book proposes a classification—detailed in Table 3.1—of the main aspects of security that combines ideas from multiple sources [22, 32, 58].

The classification consists of the following six aspects of security: *confidentiality, integrity, availability, authenticity, reliability,* and *accountability*. Starting from these core aspects, STS-ml comes with a list of security requirements, which is shown in the taxonomy of Figure 3.16.

The following sections, 3.5.1 to 3.5.6, will detail what each aspect means, explain the security requirements for each of these aspects, and provide examples that refer to the healthcare scenario.

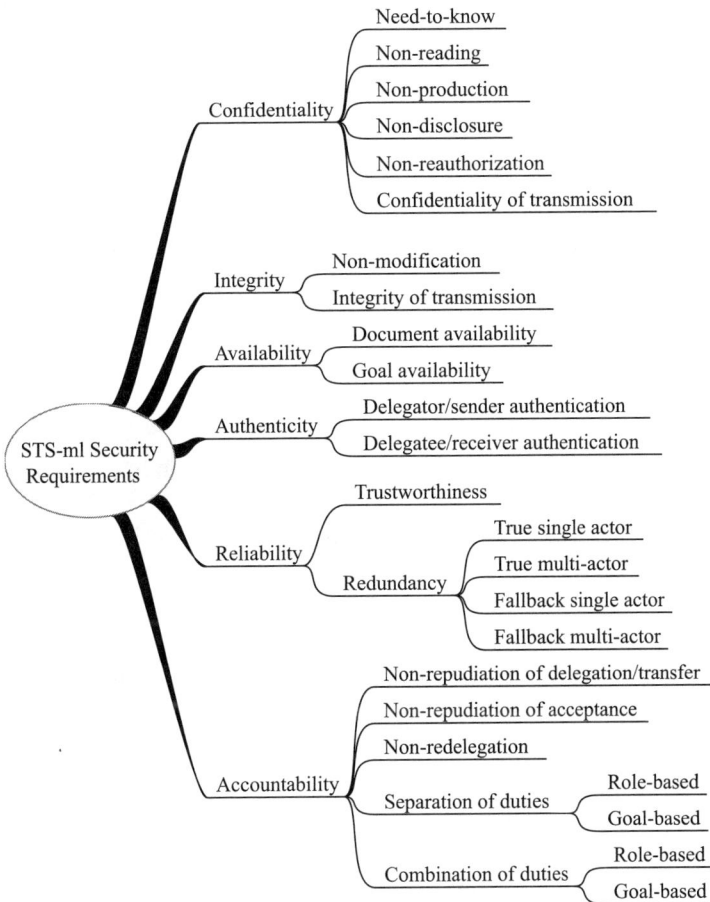

Figure 3.16
Security requirements supported by STS-ml.

3.5.1   Confidentiality

The security aspect of confidentiality encompasses both information confidentiality and privacy [58]. It ensures that private or confidential information is not made available or disclosed to unauthorized users. Moreover, it ensures that a stakeholder will specify what information related to her or him may be collected (used), and to whom that information may be disclosed.

An example of this category of requirements concerns sensitive information such as one's credit card number. Typically, one would like to ensure that only authorized people can access it (and for specific purposes), aw well as being the one who specifies these authorizations.

In STS-ml, confidentiality requirements are expressed via different types of authorizations. The language supports the following security requirements for confidentiality (illustrated in Figure 3.17):

3.5.1.1   Need-to-know

This requirement is derived from the need-to-know principle, which states that an actor shall have access to and rights about only the information necessary to accomplish its tasks. In STS-ml, this requirement is expressed by an actor when it grants an authorization to another actor, and the purpose of this authorization is not empty. In other words, the authorizer restricts the permissions granted to the authorizee to one or more goals.

The authorized actor can perform the permitted operations over documents representing that information only in the specified purpose, but not for achieving other goals. The prohibitions relate only to the specified goals as well.

Notice that when determining the purpose for which the permissions and prohibitions are specified, STS-ml considers not only the goals that are in the authorization purpose, but also all their descendants in the actor model of the authorizee. For instance, if the authorization is given on goal blood collected, and this goal is and-decomposed as in Figure 3.9, the authorization applies to the subgoals donor approved and blood examined too.

In Figure 3.17, the Patient requires the Hospital to ensure need-to-know over information health status, personal data, and medical history in the scope of goal medical advice given. This means that the Hospital can perform the allowed operations (only use) to achieve goal medical advice given (including its subgoals and descendants), but not for achieving other unrelated goals. Similarly, the prohibition to modify and disclose the same information elements is restricted to the goal medical advice given and its descendants in the actor model of the Hospital.

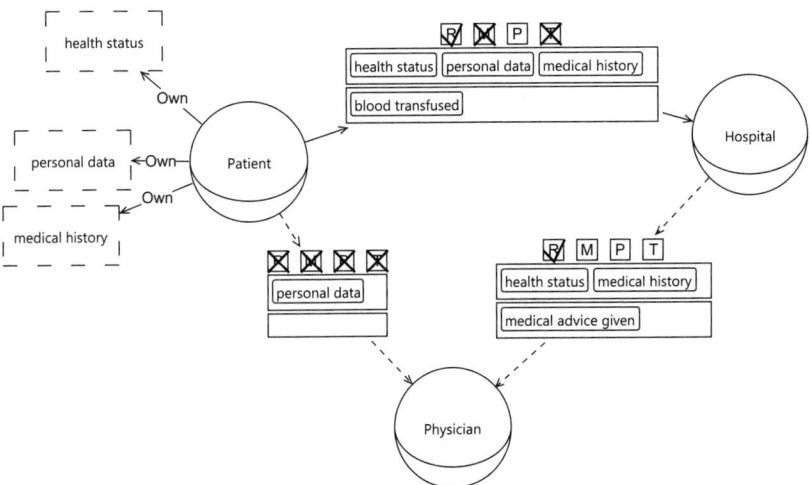

(a) Non-reading (crossed out R slot), non-production (crossed out P slot), non-disclosure (crossed out T slot), non-reauthorization (dashed line), need-to-know (goal scope slot not empty)

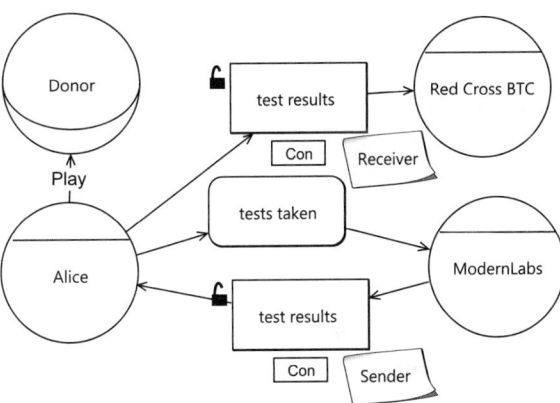

(b) Confidentiality of transmission, with sticky notes informally illustrating the required type

Figure 3.17
Confidentiality security requirements.

3.5.1.2 Non-reading

This confidentiality requirement states that the authorizer wants the authorizee not to read the information in the expressed authorization.

Non-reading is specified when the authorizer prohibits the read operation to the authorizee.

In Figure 3.17, the Patient expresses a non-reading requirement for her personal data on the Physician, because of the authorization where the Patient prohibits the usage of personal data. Notice that the very authorization also expresses other requirements that are introduced later in this section: non-disclosure, non-reauthorization, non-modification, and non-production.

3.5.1.3   Non-production

This requirement indicates that the authorizer wants the authorizee to not produce any document that contains the information in the authorization. It is expressed when the authorizer prohibits production to the authorizee.

In Figure 3.17, the Patient expresses a non-production requirement for her personal data on the Physician, by prohibiting the production operation.

3.5.1.4   Non-disclosure

The non-disclosure confidentiality requirement is expressed by an authorizer to indicate that some information should not be transmitted (disclosed) in an unauthorized way by the authorizee. The latter actor shall not transmit any document that contains the specified information to unauthorized actors.

In Figure 3.17, the Patient requires the Hospital to not disclose information health status, personal data, or medical history. Similarly, as anticipated in the section related to non-usage, the Patient requires the Physician not to disclose her personal data.

3.5.1.5   Non-reauthorization

This requirement indicates that the authorizer wants the authorizee not to redistribute the received permissions to other actors. If the authorizee receives an authorization that contains only prohibitions, then non-reauthorization cannot be specified. An authorizee is subject to this requirement in two cases:

- Explicitly, when the authorizee receives an authorization that is non-transferable (graphically, when the line connecting the authorizer and the authorizee is dashed);
- Implicitly, when no actor specifies permissions or prohibitions for performing a certain operation on given information.

The requirement has one exception: the information owner possesses all permissions, and prohibitions do not apply to such actor.

Figure 3.17 includes an example of explicit expression of this requirement: the authorization from Hospital to Physician (dashed arrow line) for information health status and medical history. Moreover, if the authorization (including only prohibitions) from Patient to Physician were removed, the scenario would include an implicit specification, because no actor would have permitted the Physician to use the Patient's personal data.

Notice the key difference between the explicit and the implicit cases. In the former case, an actor wants to ensure that another cannot do specific operations on some information. In the latter case, the lack of permissions is a temporary situation, which could be changed by any actor having permission and authority to transfer such permission, if they used this authority and specified an authorization that passed such permission. For example, in Figure 3.17, removing the authorization between Patient and Physician would allow the Hospital to pass its permission on personal data to the Physician.

Chapter 5 provides a thorough discussion on conflicting authorizations, their identification, and resolution.

3.5.1.6 Confidentiality of Transmission

This requirement indicates that the confidentiality of some information should be preserved while it is transmitted from one actor to another. In STS-ml, information is exchanged through document transmission. The burden of ensuring confidentiality of transmission may affect the sender, the receiver, or the socio-technical system's infrastructure:

- *Sender confidentiality*: the sender should ensure that the confidentiality of transmission for the given document is preserved. Thus, the requirement is expressed by the receiver to require that the sender

ensure the confidentiality of transmission of the document being transmitted. In Figure 3.17b, Alice requires ModernLabs to ensure the confidentiality of transmission of document test results.

- *Receiver confidentiality*: the receiver should ensure that the confidentiality of transmission for the given document is preserved. Thus, the
requirement indicates that the sender requires the receiver to ensure
the confidentiality of transmission of the document being transmitted.
In Figure 3.17b, Alice requires Red Cross BTC to ensure the confidentiality of transmission of document test results.

- *System confidentiality*: the system should ensure that the confidentiality of transmission of a document in transit is preserved. This indicates
that the requirement is imposed by the organization itself deeming the
document as important to preserve confidentiality while being transmitted.

Graphically, in Figure 3.17b, the transmission of document test results
from Alice to Red Cross BTC and from ModernLabs to Alice are annotated
with a padlock and with a small rectangle labeled "Con". The former
annotation indicates that a security requirement applies to the transmission (as explained later, multiple requirements can apply to the same
element). The latter annotation indicates the type of requirement: "Con"
stands for "confidentiality of transmission." Moreover, the figure also
includes sticky notes as an informal way to show the type of confidentiality of transmission that is required; note that this is not part of the
syntax of STS-ml.

3.5.2   Integrity

The security aspect of integrity ensures that information is not changed
(modified) or destroyed in an unauthorized way [58]. Notice that when
one refers to informational entities, it is very challenging to identify a
deletion, for information can be made tangible by many documents and
copies of these documents. Therefore, STS-ml focuses only on the unauthorized modification of documents that make some information tangible. Specifically, STS-ml supports the following security requirements,
illustrated in Figure 3.18.

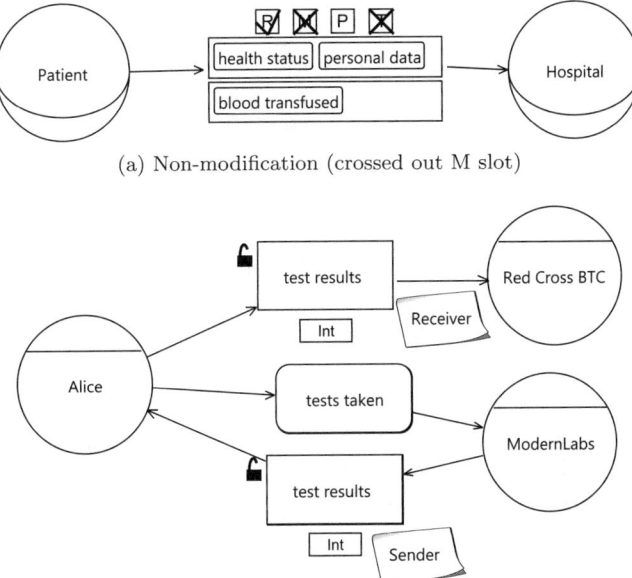

(a) Non-modification (crossed out M slot)

(b) Integrity of transmission: sticky notes informally indicate the type

Figure 3.18
Integrity security requirements.

3.5.2.1 Non-modification

This integrity requirement requires that information not be modified in an unauthorized way. The requirement is expressed through authorizations and specifies that the authorizer wants the authorizee not to modify any document that makes tangible the information in the authorization in the context of the specified purpose.

In Figure 3.18a, the Patient requires the Hospital not to modify her health status and personal data.

A non-modification requirement may seem over-restrictive: how can one model the fact that only sections of the document containing that information should not be modified? The part-of relationship within the document is the answer. One can structure a document into sub-documents (e.g., header, body, signature), each being part of the overall document. By relating information to a specific part, the other parts can be modified without affecting the integrity of the information at hand.

3.5.2.2   Integrity of transmission

This requirement indicates that some information should not get corrupted while in transit from one actor to another. In STS-ml, information is transmitted through document transmission. Integrity of transmission might be required by the sender, the receiver, and the system (meaning the socio-technical system). Therefore, the requirement is expressed as a constraint on document transmissions and is specialized into three types:

- *Sender integrity*: the sender should ensure the integrity of transmission for the given document. Thus, the requirement is expressed by the receiver to require that the sender ensure the integrity of transmission of the document being transmitted. In Figure 3.18, Alice requires ModernLabs to ensure the integrity of transmission of document test results.

- *Receiver integrity*: the receiver should ensure that the integrity of transmission for the given document is preserved. Thus, the requirement indicates that the sender requires the receiver to ensure the integrity of transmission of the document being transmitted. In Figure 3.18, Alice requires Red Cross BTC to ensure the integrity of transmission of document test results.

- *System integrity*: the system should ensure that the integrity of transmission of a document in transit is preserved. This indicates that the requirement is imposed by the organization itself deeming the document as important to preserve integrity while being transmitted.

The transmission of document test results from Alice to Red Cross BTC and that from ModernLabs to Alice are annotated with a padlock and with a small rectangle labeled "Int." The latter annotation indicates the type of requirement: "Int" stands for "Integrity of transmission."

3.5.3   Availability

Availability as a security requirement means that the system works promptly, service is not denied to authorized users, and access to and use of information is timely and reliable [58]. This aspect of security is sometimes treated separately from the others. To provide a comprehensive account on security requirements engineering, however, availability is treated as part of security requirements. STS-ml supports two types of availability: document availability and goal availability.

3.5.3.1 Document availability

This requirement indicates that the actor that possesses the document has to ensure a certain level of availability to the actor that needs the document. It is expressed over document transmissions, and it requires the sender to guarantee an availability level expressed in percentage (X%) for the document being transferred to the receiver. Availability requirements are graphically expressed by annotating document transmissions with a small rectangle labeled "Ava."

For instance, in Figure 3.19, the Hospital Authority requires Hospitals to guarantee an availability of 99.9% (the level is visualized informally as a sticky note) for the document registration record, which is created whenever a patient is hospitalized.

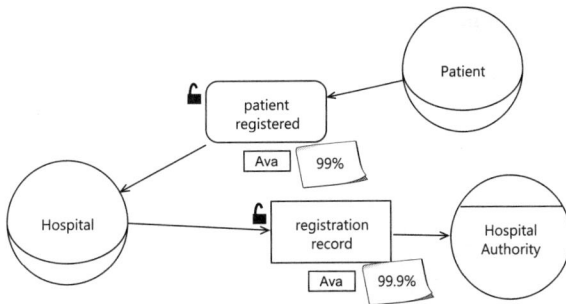

Figure 3.19
Availability security requirements, with sticky notes informally indicating the availability level.

3.5.3.2 Goal availability

This availability requirement is expressed by a delegator to indicate that a minimum level of availability should be provided by the delegatee of a given goal. Thus, it is expressed over goal delegations. Its graphical syntax is the same as for document availability.

For instance, in Figure 3.19, the Patient requests of the Hospital an availability level of 99% for the delegated goal patient registered, meaning that the Hospital should guarantee that patient registration be ensured in 99% of cases. Expressing an availability level is often required to accommodate unpredictable circumstances, such as too many patients showing up at the Hospital on the same day.

Notice that goal availability is highly related to the notion of service availability, where a provider specifies an uptime level for the service. In service-oriented settings, availability levels often become integral parts of service-level agreements between providers and consumers.

A useful tool to define availability levels, for both document and goal availability, is converting the percentage to downtime, that is, the maximum amount of time the delegatee/sender is allowed to be unavailable over a certain period (per week, month, year). Table 3.2 shows the conversion of availability percentage into downtime.

Table 3.2

Converting availability percentage to downtime.

Availability	Downtime		
	per year	per month	per week
90% ("one nine")	36.5d	72h	16.8h
95%	18.25d	36h	8.4h
98%	7.3d	14.4h	3.36h
99% ("two nines")	3.65d	7.20h	1.68h
99.5%	1.83d	3.60h	50.4m
99.8%	17.52h	86.23m	20.16m
99.9% ("three nines")	8.76h	43.8m	10.1m
99.99% ("four nines")	52.56m	4.32m	1.01m
99.999% ("five nines")	5.26m	25.9s	6.05s
99.9999% ("six nines")	31.5s	2.59s	0.605s
99.99999% ("seven nines")	3.15s	0.259s	0.0605s

3.5.4 Authenticity

The security aspect of authenticity is the property of being genuine and able to be verified and trusted [32]. Authenticity is ensured through authentication processes that aim at verifying whether users are who they say they are (entity authenticity [58]).

In STS-ml, the authenticity requirement is expressed on actors' interactions related to their assets, namely *goal delegations* and *document transmissions*. Authenticity is specialized into two variants: authentication of the delegator/sender and authentication of the delegatee/receiver.

Given a goal delegation or document transmission, one can express either delegator/sender authentication, or the delegatee/receiver version, or both. Graphically, authenticity requirements are expressed as an annotation of a delegation/transmission.

3.5.4.1   Delegator/Sender Authentication

This requirement for authenticity indicates the delegatee/receiver's request that the delegator/sender be authenticated. This is the kind of authentication that is typically implemented in electronic commerce websites, wherein a certification authority guarantees the authenticity of the seller's website.

In the healthcare scenario in Figure 3.20, the Hospital expresses the requirement that the transmission of document registration record necessitate the sender's (ModernLabs) authentication. Moreover, the delegation of goal tests taken from Alice to ModernLabs includes a delegator authentication requirement: ModernLabs wants to ensure that the test is requested by the same person who will be tested.

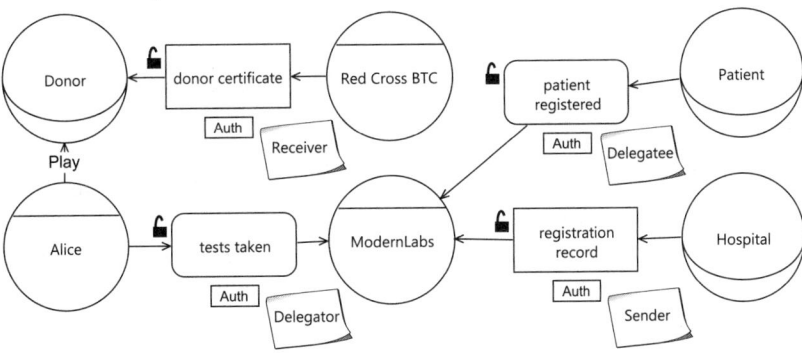

Figure 3.20
Authenticity security requirement, with informal annotations that describe the type of required authentication.

3.5.4.2   Delegatee/Receiver Authentication

This version of the authenticity requirement expresses the delegator/sender's need for the delegatee/receiver to authenticate herself. We encounter this kind of authentication every day when we browse the web and use our credentials (username/password) to access web information such as our email.

In Figure 3.20, the Patient imposes this requirement on the delegation of goal patient registered to the Hospital, for she wants to be certain that registration is performed by the Hospital she trusts. Also, the transmission of the donor certificate from Red Cross BTC to the Donor includes this requirement, which indicates that the Donor herself is the only individual who can receive the certificate.

3.5.5   Reliability

Reliability is an aspect of security that addresses the consequences of accidental errors [22]. In the age of the Internet, however, the notion of accident encompasses non-designed usages, including attackers trying to misuse the system. While reliability is sometimes treated independently from security, it is included here with the intent of providing a comprehensive approach. In STS-ml, reliability is supported through multiple requirements.

3.5.5.1   Trustworthiness

The requirement for trustworthiness is expressed over goal delegations by the delegator, which requires the delegatee to be trustworthy; that is, the goal will be delegated only to trusted delegatees. This requirement implies that the delegatee will provide a proof of trustworthiness, for example, issued by a certification authority. For instance, in the healthcare scenario (Figure 3.21), the Patient imposes a trustworthiness requirement on the Physician with respect to transf needed (performing a transfusion procedure when needed).

3.5.5.2   Redundancy

This requirement is expressed on goal delegations by the delegator, who wants the delegatee to adopt redundant strategies for a delegated goal,

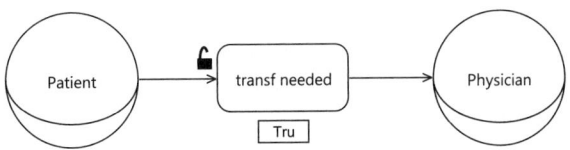

Figure 3.21
Trustworthiness security requirements.

either by using alternative internal strategies (single actor), or by relying on other actors (multi-actor). Two types of redundancy requirements are considered:

1. *Fallback redundancy*: a primary strategy should be selected to fulfill the goal, while at least one other strategy is available as backup, used only if the primary strategy fails.
2. *True redundancy*: two or more different strategies should be executed simultaneously by the delegatee.

Redundancy requirements are expressed as an annotation of a goal delegation, as shown in the healthcare scenario snippets of Figure 3.22.

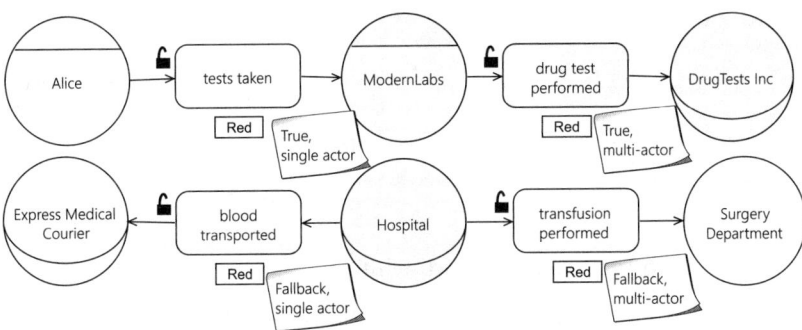

Figure 3.22
Reliability security requirements, with informal annotations that describe the type.

By intertwining single/multi-actor with the redundancy types, STS-ml supports four mutually exclusive redundancy security requirements:

1. *True redundancy single actor* indicates that one actor has to deploy concurrent means for achieving the delegated goal. For example, in

Figure 3.22, Alice requires ModernLabs to provide true redundancy for goal tests taken, for example, testing the blood through two instruments.

2. *True redundancy multi-actor* specifies that the delegatee shall ensure that more than one actor (possibly including the delegatee herself) shall act concurrently toward the achievement of the delegated goal. In Figure 3.22, ModernLabs requires DrugTests Inc to use concurrent ways of performing drug tests, with multiple actors involved.

3. *Fallback redundancy single actor* means that the same actor has to provide a fallback solution, in case the original one fails. In Figure 3.22, the Hospital requires the Express Medical Courier to deploy a fallback solution for goal blood transported. A possible way to do so is to reserve a backup van, which is used only if the designated one is not working.

4. *Fallback redundancy multi-actor* expresses that the delegatee should ensure the involvement of multiple actors toward the achievement of the delegated goal. In Figure 3.22, the Hospital specifies this requirement to the Surgery Department for goal transfusion performed. This means that more than one physician should be ready to perform the transfusion: one is the designated surgeon, the other ones are backup options, should the first one become unavailable.

3.5.6   Accountability

This security aspect refers to the requirements for actions of an entity to be traced uniquely to that entity [32]. STS-ml supports expressing accountability security requirements in different ways, as shown in the following subsections.

3.5.6.1   Non-repudiation

A key requirement related to accountability is non-repudiation, that is, preventing either the sender or the receiver from denying a transmitted message. In STS-ml this requirement is expressed not only with respect to transmitting messages (document transmission), but also with respect to goal delegation (which transfers responsibility for goal fulfillment).

This requirement is specified to prevent the two interacting actors from denying having transferred the document/delegated the goal or received the document/accepted the delegation respectively. Therefore, non-repudiation is specialized into two types:

- *Non-repudiation of acceptance*: this indicates that the delegator/sender requires the delegatee/receiver not to repudiate the delegation/transfer of a goal/document. In Figure 3.23, the Patient requires the Hospital not to repudiate the acceptance of the delegation of goal patient registered.

- *Non-repudiation of delegation/transmission*: this is expressed by the delegatee/receiver to require that the delegator/sender not repudiate the delegation/transmission of the goal/document. In Figure 3.23, ModernLabs requires Alice not to repudiate that she has delegated goal tests taken.

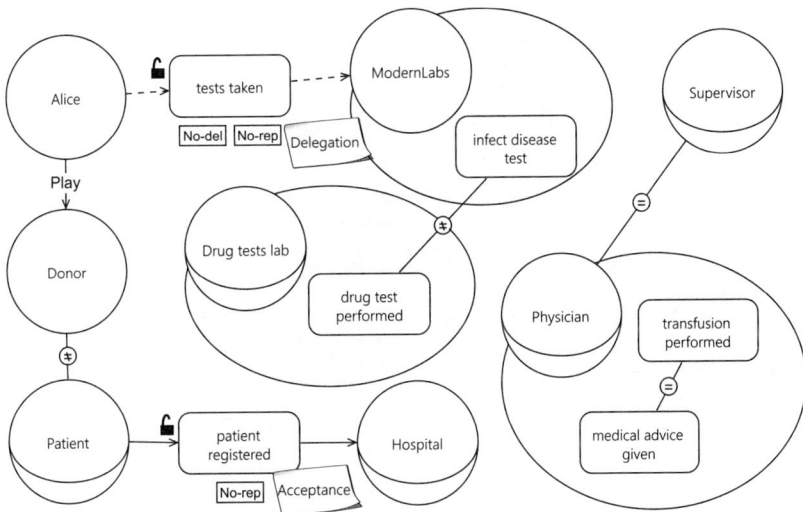

Figure 3.23
Accountability security requirements, with sticky notes informally denoting the accountability type.

3.5.6.2 Non-redelegation

This requirement is expressed over goal delegations, and it is the delegator's request that the delegatee take full responsibility for achieving the delegated goal, without relying on any other actor. The delegatee should therefore avoid delegating the goal, or subgoals, if any exist. Graphically,

the requirement is represented using two means: an annotation for delegations with label "No-del", and a dashed line, for consistency with the
non-reauthorization notation (see Section 3.5.1.5).

A main reason for specifying a non-redelegation requirement concerns
trust: the delegator trusts that specific delegatee for some goal and does
not trust other actors the delegatee might want to involve. In Figure
3.23, Alice requires ModernLabs not to redelegate goal tests taken to other
actors, such as third-party labs or technicians.

3.5.6.3   Separation of duties

This accountability requirement indicates segregation of duties among
different people, especially when dealing with critical tasks. An example
of this requirement is the procedure for opening a bank's safe deposit
box, which often requires the joint effort (and information) of more than
one person.

Graphically (see Figure 3.23), this is represented as an arrow between
two entities annotated with the "not-equal" ($\neq$) symbol. Specifically,
separation of duties (SoD) comes in two versions in STS-ml:

- *Role-based SoD* defines two roles as incompatible; that is, when specified between the two roles, it does not allow the same agent to play
  both the roles. For example, in Figure 3.23, roles Patient and Donor
  are incompatible: if Alice is a Donor, she cannot be a Patient (blood
  receiver) at the same time.

- *Goal-based SoD* defines incompatible goals; that is, an actor should
  not pursue and achieve both goals among which SoD is defined. For
  example, in Figure 3.23, ModernLabs' goal to test for infectious dis-
  eases conflicts with DrugTests Inc's goal of performing drug tests. This
  requires that different actors should be responsible for these two tests.

3.5.6.4   Combination of duties

This requirement—sometimes referred to as *retain familiar* [50]—is symmetric to separation of duties. It expresses that some entities be ascribed
to the same actor. Graphically (see Figure 3.23), this is represented as
an arrow between two entities annotated with the "equal" ($=$) symbol.
Combination of duties (CoD) comes in two versions in STS-ml:

- *Role-based CoD* defines a binding between roles; that is, if an agent adopts either of the roles, it also has to adopt (play) the other role. In Figure 3.23, a role-based combination of duties requirement is expressed over roles Supervisor and Physician; that is, a Physician also has to supervise trainees in the hospital.

- *Goal-based CoD* defines a binding between goals; that is, if an agent achieves one of the goals among which CoD is defined, it should achieve the other goal too. In Figure 3.23, the Physician role includes a goal-based CoD for goals transfusion performed and medical advice given. This means that the same Physician has to be responsible for both goals (e.g., in order to increase the patient's trust about the physician/hospital).

3.6 Chapter summary

In this chapter, the modeling primitives of STS-ml are presented. These constructs allow for expressing the stakeholders in a socio-technical systems, their objectives, and the security requirements they want other stakeholders to comply with.

The supported security requirements of the language are shown and classified according to a taxonomy of security aspects, which consists of confidentiality, integrity, availability, authenticity, reliability, and accountability.

The language complies with ten important principles that characterize security requirements in socio-technical systems:

1. A socio-technical perspective is provided by modeling actors (social and technical) that interact via goal delegation and document transmission.
2. Security is imposed as constraints that affect different types of interactions: goal delegations, document transmission, and authorizations.
3. STS-ml relies on high-level assets, by modeling actors' goals, information, and documents.
4. Representing events and how these threaten supporting assets is an essential part of the language.
5. The existence of multiple stakeholders is acknowledged by enabling every stakeholder to express requirements.
6. STS-ml models are diagrammatic, and they come with a formal semantics (which will be shown in Chapter 5).

7. The supported security requirements are classified and presented following a taxonomy derived from existing standards.

8. Minimality of concepts is ensured by carefully avoiding overlaps between the entities and relationships.

9. Security requirements can be traced back to their requester and to the motivation/goal that originates the requirement.

10. The language focuses on security needs, and not on the mechanisms for fulfilling these needs.

This chapter provides the foundations for the rest of the book, which uses the modeling primitives described here. The reader should make sure that the proposed concepts are well understood before moving on to the next chapter, which explains how these primitives are combined into three types of views.

3.7 Exercises

Review questions

Q3.1. What is a socio-technical system? How does it differ from other types of systems? List a few examples.

Q3.2. What is the difference between a role and an agent?

Q3.3. Define assets. What is the difference between informational and intentional assets?

Q3.4. How are threats represented in STS-ml?

Q3.5. List the main differences between information and document.

Q3.6. Explain the part-of relationship, also detailing which elements it can relate in STS-ml.

Q3.7. Describe the security requirements types that STS-ml supports.

Q3.8. Explain the difference between delegator and sender authentication.

Problems

P3.1. Consider the financial department of a telecommunications company. Identify the actors involved in such an environment and classify them into roles and agents.

P3.2. Consider a scenario in which a citizen (say, Charles Foster Kane) wants to buy a land lot through a real estate agency. He wants the lot to be in a residential area and has set an upper price limit. He registers with the agency to check the various offers and eventually selects one. The final evaluation of the lot, however, is performed by a civil engineer to estimate the construction capabilities. Complete the following tasks:

a. Define the actors in this scenario.
b. Identify Charles' objectives (goals).
c. Propose alternative ways for Charles to achieve his goals.
d. Explain whether Charles interacts with other parties to achieve his goals. If so, how?

P3.3. Take the land-buying scenario from Problem P3.2. Think of the information that Charles has and wants to protect. Identify the information that Charles exchanges with other actors.

P3.4. Consider again the scenario from Problem P3.2. What are Charles' main security concerns? Give examples of security requirements for the interactions he participates in.

P3.5. Consider a travel agency service that offers customers the possibility to check various destinations, book flights and hotels and so forth. Give examples of confidentiality, integrity and availability requirements related to this system considering a customer using this service. Identify at least one requirement per category.

4 Social, Information, and Authorization Views

This chapter shows how the concepts and relationships described in Chapter 3 can be used in conjunction to build STS-ml models, which are used for expressing the security requirements for the system under design.

Specifically, the reader is provided with an explanation on how these primitives are combined into three complementary views, namely the *social*, the *information*, and the *authorization* view. Roughly, each of these views presents a different perspective on the system-to-be, which highlights specific aspects and is suitable for certain stakeholders. The overall STS-ml model for the system is the union of the three views.

After arguing for the need and showing the benefits of a multi-view approach to the creation of STS-ml models (Section 4.1), Sections 4.2 to 4.4 explain how to construct each of the views to describe the stakeholders' security needs that justify the security requirements for the system-to-be.

4.1 Multi-view modeling in STS-ml

Many of the models that humans create tend to become large and complicated, due to the inherent complexity of the represented domains. Think, for instance, of models describing power plants, cities, electronic boards, or networks. Socio-technical systems are not an exception, as they describe the goals of and interactions among people, organizations, and technical systems.

A useful way to tackle such complexity is by acknowledging that these models represent multiple perspectives, each of which highlights a specific aspect of the system-to-be. Each of these perspectives represents a *view* on the overall model. For example, possible views for the design model of a house represent the plumbing system, the electrical system, and so on. Similarly, geographic information systems have views called "layers" to separate the different aspect of a map, such as water, vegetation, buildings, and altitude.

STS-ml employs multi-view modeling by arranging the concepts in Chapters 3 into three different yet complementary views that together represent the overall STS-ml model. The advantage of this choice is that analysts can examine one specific aspect of the system at a time. Clearly, these views are interrelated—they describe facets of the *same* system—and changes in one view may require modifications in the other views.

Based on these premises, STS-ml modeling is performed by focusing on three views, each representing a specific perspective of the system. Together, these views provide a comprehensive representation of the important security dimensions of a socio-technical system:

- *Social view*, which captures *social* and *organizational* aspects, in which actors are modeled along with their objectives and interactions with others.

- *Information view*, to represent informational assets, that is, information actors own and want to protect.

- *Authorization view*, to model the flow of permissions and prohibitions among the actors that participate in the system.

This separation of concerns facilitates the collaboration between requirements analysts and security engineers, allowing each of them to focus on modeling the aspects related to their expertise. For instance, a requirements analyst would typically contribute mostly with information regarding the social and organizational aspects (social view), while a security engineer would mainly model the permissions and prohibitions among actors (authorization view).

The following sections describe which modeling primitives are used to create an STS-ml model through each view and show the resulting models from the modeling of each view.

4.2   Social view

The *social view* represents the stakeholders of the system, their own goals, and the interactions among them. Stakeholders are represented via actors that are intentional (they have objectives they aim to attain) and social (they interact with others to achieve their respective objectives).

Therefore, the social view allows representing actors together with their intentional assets (their *goals*) and documents, the representation of which results in constructing actor models (see Section 3.2.3). Apart from representing the internal rationale of the various identified actors, the social view enables capturing the interactions among these actors, either to fulfill their objectives (by *delegating goals*) or to obtain information (by *transmitting documents*).

These concepts and relationships have already been explained in Chapter 3; here, only the supported concepts and relationships that are used in the social view are presented, through the following four tables:

- Table 4.1 and Table 4.2 present the main concepts and intentional relationships.

- Table 4.3 presents the social relationships supported by the social view, namely *play, goal delegation,* and *document transmission.*

- Table 4.4 presents the *event* concept and the *threaten* relationship, which support threat modeling.

Tables 4.1–4.8, for illustrative purposes, use interchangeably the graphical syntax of role and agent to visualize an actor, for these relationships apply to both roles and agents. The *play* relationship is the exception, as in this case the difference between agents and roles is crisp and necessary.

Figure 4.1 depicts a partial social view for the healthcare scenario of Section 1.5. In the following sections, chunks of this model are used to explain and illustrate the concepts and relationships summarized in Tables 4.1–4.4.

4.2.1   Concepts and intentional relationships

The participating stakeholders are modeled via roles and agents (Table 4.1). In the healthcare scenario, various stakeholders are identified,[1], including Alice, donors, patients, hospitals, physicians, laboratories (ModernLabs and Drug Test Inc), and Red Cross BTC.

Donors, patients, hospitals, and physicians are modeled as roles, while Alice, ModernLabs, Drug Test Inc, and Red Cross BTC are represented through the concept of agent (see Figure 4.1). The reason for this is that *roles* refer to general actors that are instantiated at runtime—that is, they are adopted (played) by concrete participants in the socio-technical system (here, the healthcare system)—while *agents* refer to concrete

---

1 While stakeholders are introduced via roles and agents in the social view, these concepts could be introduced in any of the supported views.

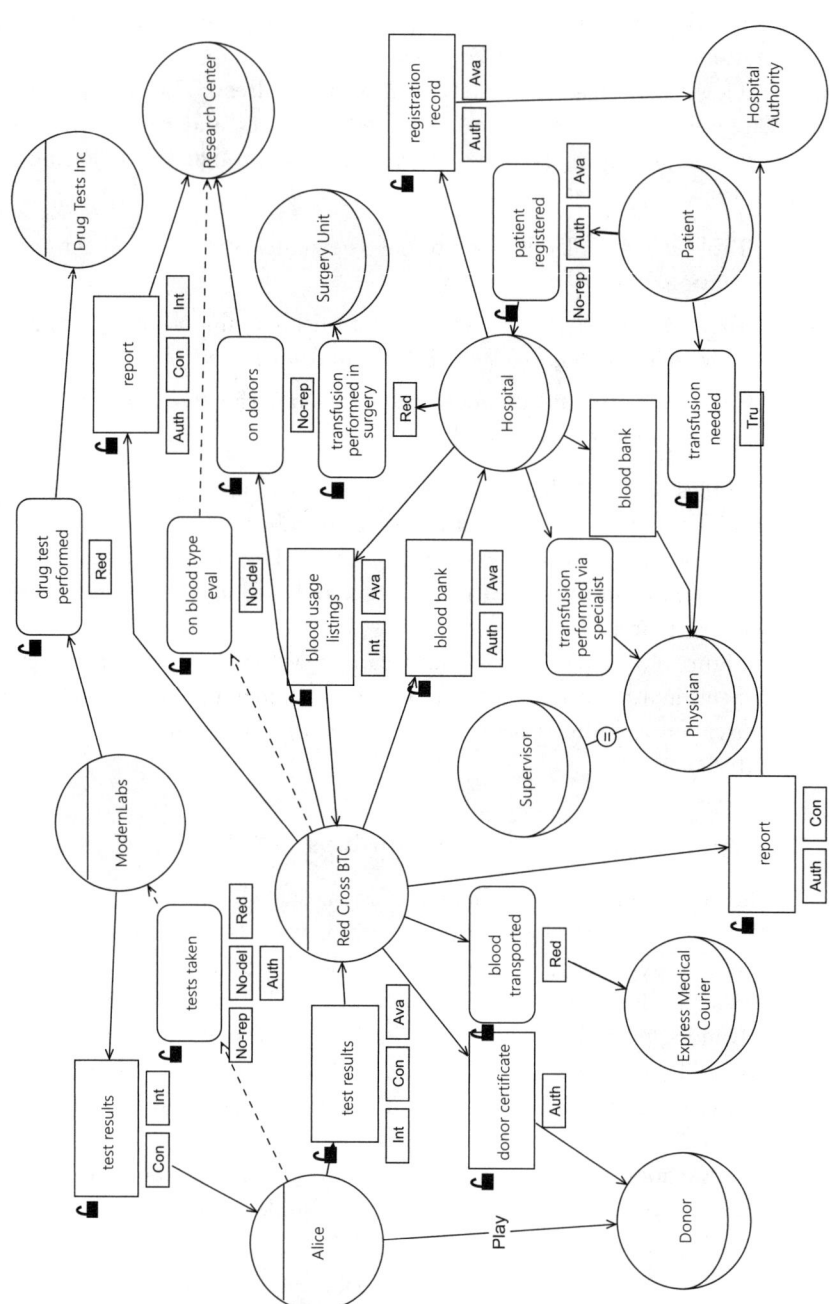

Figure 4.1

Partial STS-ml social view of the healthcare scenario: interactions among actors.

Table 4.1

Social view: concepts.

Graphical notation	Syntax and description
R	role(R): an abstract characterization used to model a class of participants, defining a set of responsibilities for said participants (e.g., professor, student)
A	agent(A): a concrete participant known to be in the system already at design time (e.g., John, Laura)
G	goal(G): stakeholders' objectives (e.g., exam taken)
D	document(D): transferable information (e.g., transcripts file)

Table 4.2

Social view: intentional relationships.

Graphical notation	Syntax and description
	wants(A, G): actor A wants to achieve goal G; models the actor's intention to achieve the goal (e.g., student wants to pass the exam)
	possesses(A, D): actor A possesses document D; models the possession of the document by an actor (e.g., professor has the exam correction sheet)
	reads/modifies/produces (A, G, D): actor A reads, modifies, and produces document D when fulfilling goal G (e.g., professor reads student exams to grade them)
	decomposes(A, G, S, DecT): A decomposes goal G into the subgoals in set S, where $S=\{G_1,\ldots,G_n\}$ and $n \geq 2$, and the decomposition is of type DecT, such that DecT $\in \{$and, or$\}$ (e.g., a student passes the exam if she sits in for the test and obtains a sufficient grade)

entities participating in the system, already known by the requirements analyst at design time. For example, the identity of the particular donors is unknown, but the responsibilities encapsulated in that *role* are known. On the other hand, some agents are known: Alice intends to become a donor, ModernLabs is the laboratory in charge of infectious disease tests, and so on.

The social view supports the modeling of stakeholders' *goals* (see Table 4.1). In the healthcare scenario, the various modeled roles and agents are analyzed to adequately represent their goals and how they intend to pursue the identified goals. For instance, in Figure 4.2, physicians (role Physician) intend to give medical advice to patients (goal medical advice given). To pursue this goal, physicians need to consult patients' medical records and perform a medical exam to assess the current medical status. Therefore, goal medical advice given is *and-decomposed* into goal medical record consulted and goal medical exam performed.

Patient has the goal of receiving medical treatment (treatment received). In the context of the blood transfusion service, medical treatment is considered to be about blood transfusion. Therefore, the patient needs to get registered to receive treatment and the transfusion: goal treatment received is *and-decomposed* into goals patient registered and transfusion needed.

Alice has the goal blood donated, which is *and-decomposed* into tests taken and neg results received (see Figure 4.3). Donor has the goal of donating blood regularly (blood donated regularly). Red Cross BTC has goal blood distributed, which is *and-decomposed* into goals blood collected, blood consumption estimated, and blood transported; goal blood collected is further *and-decomposed* into goals blood examined and donor approved, while goal blood consumption estimated is *and-decomposed* into goals stat analysis performed and blood usage evaluated; finally, goal stat analysis performed is *or-decomposed* into on blood type eval, on hospital requests, and on donor, to indicate alternative types of statistical analyses. A similar interpretation applies to the various goals and goal decompositions for each of the modeled actors in the social view of the healthcare scenario of Section 1.5.

Apart from actors' goals, the social view also features the documents that actors possess and might need to manipulate to achieve their respective goals. In Figure 4.2, Patient possesses document medical record and reads this document to get registered at the hospital (i.e., to achieve goal patient registered). Physician modifies document blood bank when transfusion is needed (goal transfusion needed). In Figure 4.3, Donor needs a donor

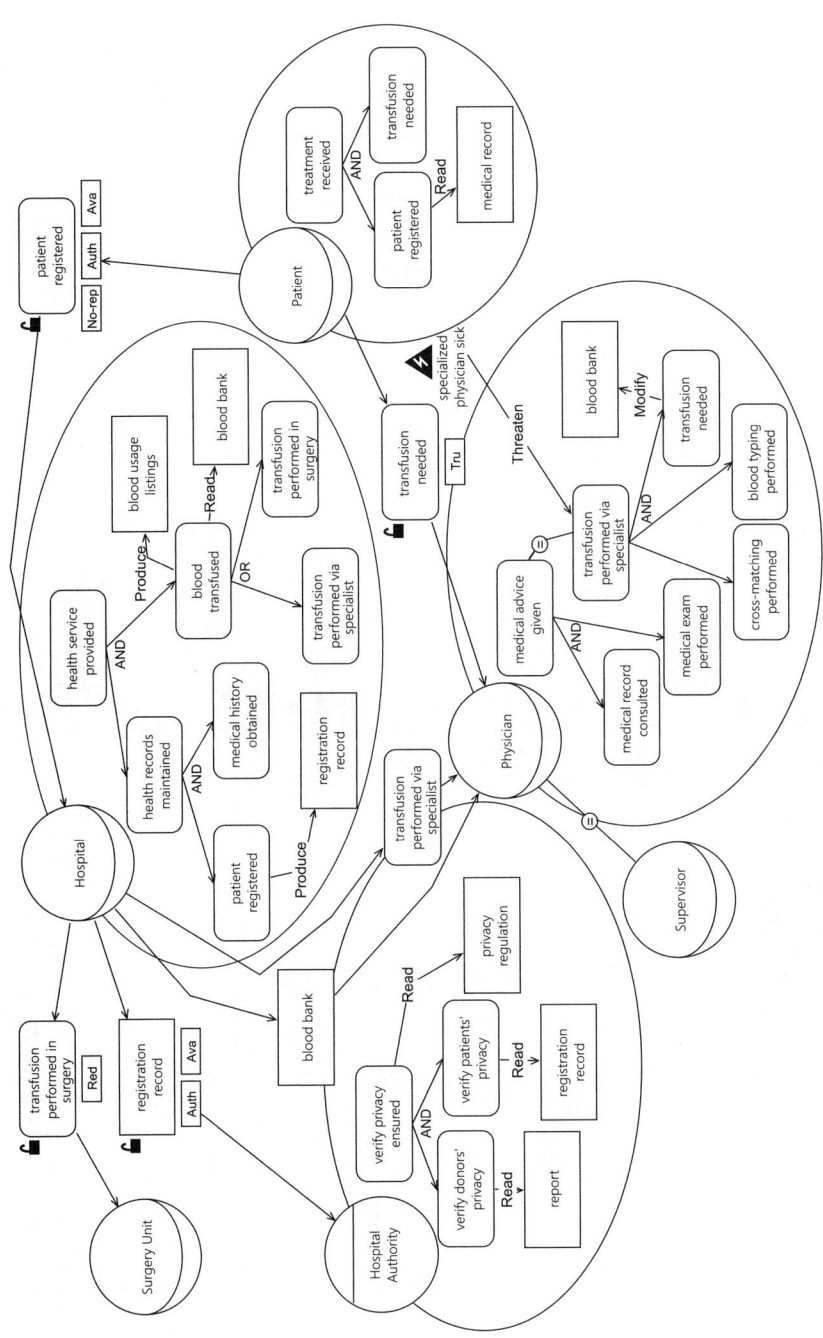

Figure 4.2
STS-ml social view of the healthcare scenario, part 1.

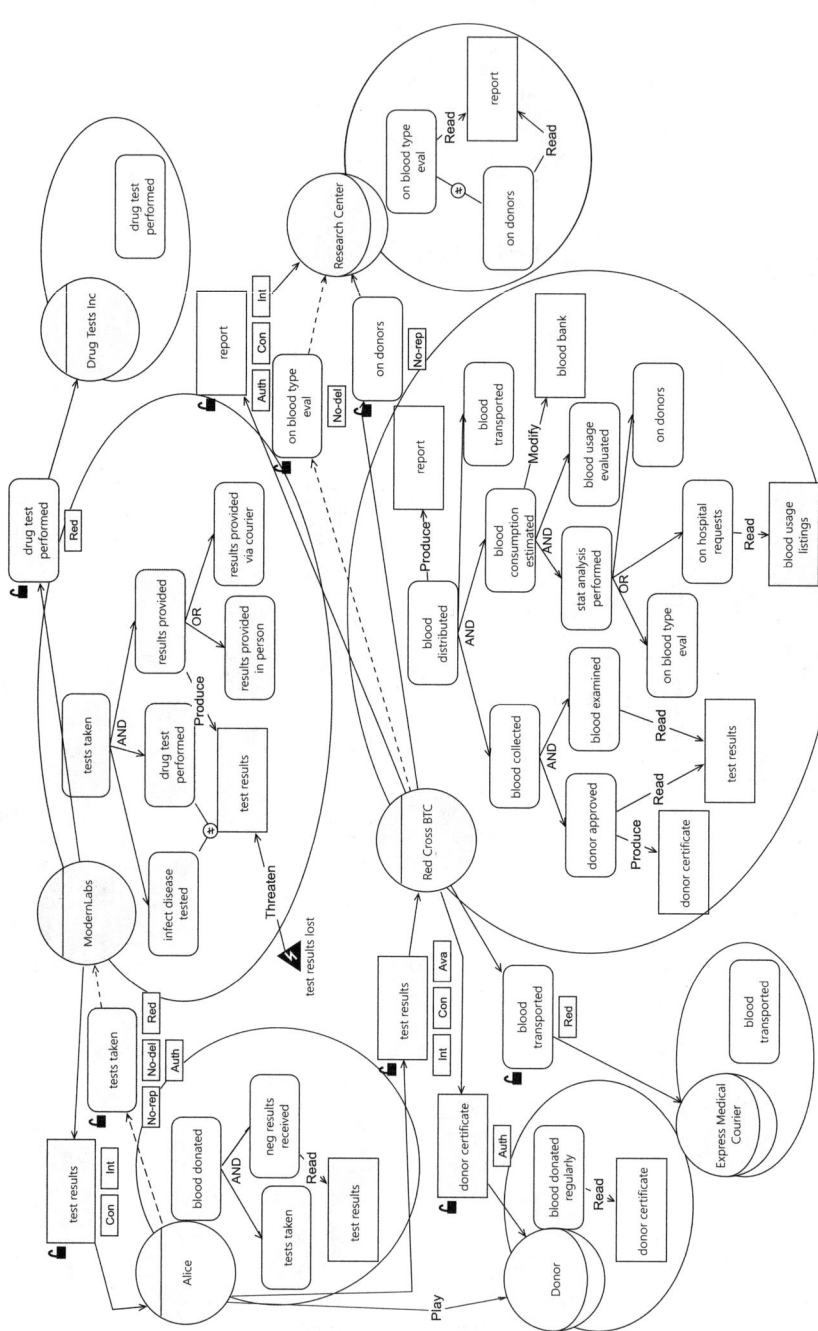

Figure 4.3
STS-ml social view of the healthcare scenario, part 2.

certificate to donate blood regularly, which is represented through the read intentional relationship between goal blood donated regularly and document donor certificate. Alice reads document test results to verify whether her blood is approved to be donated based on the test results. Red Cross BTC produces document report for goal blood distributed, modifies document blood bank to estimate blood consumption (goal blood consumption estimated), reads document test results to approve donors, produces document donor certificate for the same goal (donor approved), and finally reads document blood usage listings to perform statistical analysis on hospital requests (goal on hospital requests).

4.2.2 Social relationships

The social view supports the modeling of social relationships between actors in the given socio-technical system. Specifically, this view supports three social relationships: *play*, *goal delegation*, and *document transmission* (see Table 4.3).

Table 4.3

Social view: social relationships.

Graphical notation	Syntax and description
	plays(Ag_1, R_2): models the adoption of roles by agents; i.e., agent Ag_1 plays role R_2 (e.g., John plays role professor)
	delegates(A_1, A_2, G): models the transfer of responsibilities from an actor to another; i.e., actor A_1 (delegator) delegates the fulfillment of goal G (delegatum) to actor A_2 (delegatee) (e.g., professor delegates grading exams to teaching assistant)
	transmits(A_1, A_2, D): specifies the exchange of documents between two actors; i.e., actor A_1 (sender) transmits document D to actor A_2 (receiver) (e.g., teaching assistant transmits exam results to professor)

In the healthcare scenario, it is known that Alice wants to become a donor; therefore, this is represented through the *play* relationship between the agent Alice and the role Donor. No information exists about other actual participants of the healthcare socio-technical system, and hence this is the only play relationship represented for this scenario.

An example of *goal delegation* is that of Patient delegating goal transfusion needed to Physician. This delegation results in the second actor having the goal—that is, Physician has goal transfusion needed—which is part of the transfusion procedure performed by a specialized physician (see Figure 4.2). The Hospital delegates goal transfusion performed via specialist to the Physician. Alice relies on ModernLabs to take tests, delegating goal tests taken (Figure 4.3). ModernLabs delegates goal drug test performed to Drug Tests Inc, which is a laboratory specialized in drug tests. Similar analyses apply to the rest of the goal delegations represented in Figure 4.1.

Note that in STS-ml, only leaf goals (not decomposed) can be delegated. To express that a goal can be achieved either by the delegator actor decomposing it or by delegating to another actor, one can or-decompose the goal into two subgoals, the former being further decomposed to represent the internal achievement, the latter being delegated. This convention makes it clear which actor is responsible to pursue which goal.

As far as *document transmissions* are concerned, Figure 4.1 shows how ModernLabs transmits the document test results to Alice, who further transmits the same document to Red Cross BTC for evaluation and approval. Also, Red Cross BTC transmits document donor certificate to Donor, while it transmits blood bank to Hospital. The latter transmits document blood usage listings to Red Cross BTC, which needs this information to estimate blood consumption. More examples can be found in Figure 4.1.

4.2.3   Events and threats

The social view supports representing events threatening the assets of the stakeholders in the system under design. Therefore, the entity *event* and the relationship *threaten* are part of the social view (see Table 4.4). For instance, in the healthcare scenario, the event specialized physician sick threatens goal transfusion performed via specialist of Physician (see Figure 4.2), while the event test results lost threatens document test results produced by ModernLabs (see Figure 4.3).

4.2.4   Security requirements in the social view

The social view enables expressing security requirements over the social relationships that take place among the modeled actors. In the following paragraphs, the security requirements supported by the social view are listed, grouped by social relationship types.

Table 4.4

Social view: events and threats.

Graphical notation	Syntax and description
	event(E): represents uncertain circumstances that affect actors' assets (e.g., forgetting an important document in the tray of a shared printer)
	threaten(E, wants(A, G)) or threaten(E, possesses(A, D)): models the influence of an event over actors' supporting assets; i.e., event E threatens goal G or document D of actor A (e.g., disclosing confidential information to unauthorized users)

Given that the supported security requirements have been extensively explained in Section 3.5, this section provides only the list of security requirements types that can be captured in the social view, while providing an example where applicable to illustrate these security requirements types with the help of the motivating scenario (Section 1.5).

Over goal delegations. Security requirements are specified over goal delegations by selecting the applicable security requirements types. To do so, delegations are annotated via security requirements types (representing security needs) that the interacting actors want each other to comply with. The selection of at least one security need shows a padlock on the goal (see Table 4.5). That is, a closed padlock sign means that one or more security needs are specified over the goal delegation. The selection of security needs is graphically represented explicitly through an open padlock sign and small labeled boxes below the delegated goal; each box has a different label.

The following security requirement types are expressed in the social view over goal delegations:

1. *Non-repudiation of delegation/acceptance*: for instance, actor Modern-Labs requires Alice's non-repudiation of the delegation of goal tests taken, while the Patient requires the Hospital's non-repudiation of acceptance when delegating goal patient registered (see Figure 4.1).
2. *Redundancy*, in its four mutually exclusive types: *true redundancy single, fallback redundancy single, true redundancy multi*, and *fallback*

Table 4.5

Delegations and the representation of security requirements types.

Graphical notation	Syntax and description
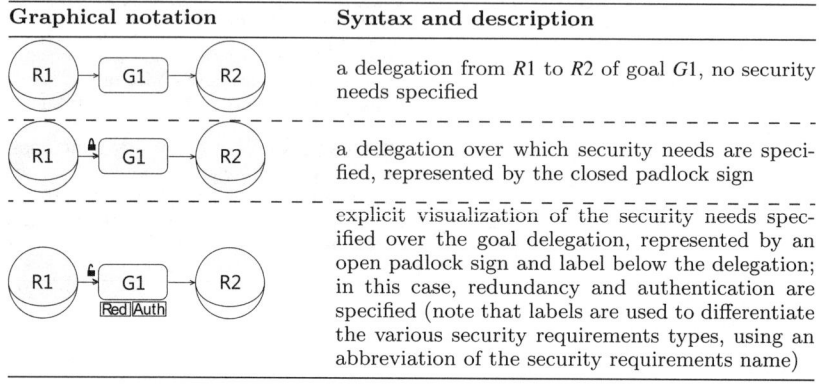	a delegation from $R1$ to $R2$ of goal $G1$, no security needs specified
	a delegation over which security needs are specified, represented by the closed padlock sign
	explicit visualization of the security needs specified over the goal delegation, represented by an open padlock sign and label below the delegation; in this case, redundancy and authentication are specified (note that labels are used to differentiate the various security requirements types, using an abbreviation of the security requirements name)

redundancy multi (see Section 3.5.5.2).[2] In Figure 4.1, Alice requires true redundancy single for goal tests taken; the Express Medical Courier requires the Hospital to employ fallback redundancy single for goal blood transported; ModernLabs requires of Drug Tests Inc true redundancy multi for goal drug test performed; while the Hospital requires of the Surgery Unit fallback redundancy multi for goal transfusion performed.

3. *Non-redelegation*: for instance, Red Cross BTC requires Research Center not to redelegate statistical analysis on donors.

4. *Trustworthiness*: in the example, the delegation of goal transfusion needed from Patient to Physician can take place only when trustworthy physicians are available (see Figure 4.2).

5. *Goal availability*: for instance, the Patient request from the Hospital an availability level of 99% for the delegated goal patient registered, meaning that the hospital should guarantee that patient registration will be ensured in 99% of cases. As we show in Chapter 8, the specific availability level (not visualized in the figure) is expressed through the software tool (Chapter 7) that accompanies the presented approach.

6. *Delegator/delegatee authentication*: for example, the delegation of goal tests taken from Alice to ModernLabs includes a delegator authentication

2 Note that mutually exclusive security requirements types are graphically represented with the same label, "Red." Details of the type specification through the supporting toolset (Chapter 7) are provided in Chapter 8.

requirement (Figure 4.3). The delegation of goal patient registered from Patient to Hospital, on the other hand, includes a security requirement concerning the delegatee (the Hospital) authentication (Figure 4.2).

Over document transmissions. Security requirements over document transmissions are specified in a similar way to those over goal delegations. That is, document transmissions are annotated via security requirements types (representing security needs) that the interacting actors want each other to comply with. The selection of at least one security need shows a padlock on the document (see Table 4.6). That is, a closed padlock sign means that one or more security needs are specified over the document transmission. The selected security needs are shown explicitly through an open padlock sign and small labeled boxes below the transmitted document; each box has a different label.

Table 4.6

Document transmissions and the representation of security requirements types.

Graphical notation	Syntax and description
	a document transmission from $R1$ to $R2$ of document $D1$, no security needs specified
	a document transmission over which security needs are specified, closed padlock sign
	explicit visualization of security needs specified over the document transmission, represented by an open padlock sign and labels for security needs below the transmission. In this case, integrity and confidentiality are specified. Note that labels are used to differentiate the various security requirements types, and thus they have different acronyms, the latter being an abbreviation of the security requirements name

The following security requirement types constrain how documents are transmitted among actors:

1. *Non-repudiation of transmission/acceptance*: for instance, the Hospital requires the Red Cross BTC not to repudiate transmission of document blood bank, as well as the acceptance of the transmission of document blood usage listings, see Figure 4.1.

2. *Integrity of transmission*, in its three forms: *sender integrity, receiver integrity*, and *system integrity* [3] (see Section 3.5.2.2). For instance, in Figure 4.1, the Hospital Authority requires the Hospital to ensure sender integrity of document registration record, Red Cross BTC requires Research Center receiver integrity over the transmission of report, while a system integrity requirement is specified over the transmission of document report from Red Cross BTC to Hospital Authority.

3. *Document availability*: for instance, the Hospital requires an availability level of 99.9% for the document blood bank from Red Cross BTC.

4. *Sender/receiver authentication*: in Figure 4.1, the transmission of document test results from Alice to Red Cross BTC requires sender authentication, while the transmission of document donor certificate from Red Cross BTC to Donor requires receiver authentication.

5. *Confidentiality of transmission*, in its three forms (see Section 3.5.1.6): *sender confidentiality, receiver confidentiality*, and *system confidentiality*.[4] For instance, the transmission of document test results from Alice to Red Cross BTC requires sender confidentiality; the transmission of document test results from ModernLabs to Alice requires receiver confidentiality; while the transmission of the document report from Red Cross BTC to Research Center requires system confidentiality.

Over responsibility uptake. The security requirements to constrain the uptake of responsibilities, that is, the adoption of roles and the pursuit of goals, are described in the following paragraphs. Remember that these security requirements types are derived from organizational constraints that are translated to a set of relationships, *incompatible* (represented as a circle with the unequal sign within) and *combines* (represented as a circle with the equals sign within).

1. *Separation of duties* (SoD), over roles or goals, *role-based SoD* and *goal-based SoD*. For instance, there is a separation of duties between the goals infect disease tested and drug test performed, which are defined as incompatible (see Figure 4.3). There is no example of separation of duties between roles in the healthcare scenario.

---

3 Note that these mutually exclusive security requirements types are graphically represented with the same label "Int". Details of the type specification through the supporting toolset (Chapter 7) are provided in Chapter 8.

4 These are not graphically shown distinctively; rather the same label "Con" is used.

2. *Combination of duties* (CoD), over roles or goals, *role-based CoD* and *goal-based CoD*. For instance, there is a combination of duties between roles Physician and Supervisor, while there is a goal-based combination of duties for the agent responsible for goals medical advice given and transfusion performed via specialist (Figure 4.2).

4.3 Information view

As explained in Section 3.2.2.1, STS-ml distinguishes between *information* (the data that actors own, care about, and may deem confidential) and its representation via *documents*. The latter, intended in a broad sense (e.g., an email or a text message are documents too), are the exchangeable means actors use to transfer information with each other.

While the social view includes documents and their transmission, it does not account for the informational content of the exchanged documents. This is useful in security requirements analysis to determine, for instance, whether information was exchanged among authorized users.

The *information view* allows representing the informational content of the documents modeled in the social view. The contents of the view indicate information entities and their owners, and provide a structured representation of both information entities and documents. The supported concepts and relationships are listed in Table 4.7.

Figure 4.4 represents a partial information view for the healthcare motivating scenario. The social view (Figure 4.2) shows that Patient possesses document medical record, for that document is within the scope of that actor, and no other actor is transferring medical record to Patient. In the information view, the analyst defines what information is contained in this document. Figure 4.4 depicts how medical record makes tangible the information personal data, which is owned by the Patient. The Patient also owns the information medical history. Similarly, Alice owns her personal information and health status, the Hospital owns information blood needs, while Red Cross BTC owns blood info.

Information can be represented by one or more documents (through multiple Tangible By relationships). For instance, the information personal information owned by Alice is made tangible by both the document health record and the document donor certificate. Also, one or more information entities can be made tangible by the same document. For instance, health status and personal information are both made tangible by document test results of ModernLabs.

Table 4.7

Information view: concepts and relationships.

Graphical notation	Syntax and description
I_1	information(I_1): informational entities (e.g., name, student grade)
A_1 —Own— I_2	owns(A_1, I_2): actor A_1 is the legitimate owner of information I_2; i.e., it has full rights over that information (e.g., students own their personal information)
D_2 Tangible By I_1	makes-tangible(I_1, D_2): document D_2 materializes information I_1 (e.g., transcripts materialize information about course results)
I_2 Part Of I_1	part-of-i(I_1, I_2): information I_1 is part of information I_2 (e.g., course description is part of course syllabus)
D_2 Part Of D_1	part-of-d(D_1, D_2): document D_1 is part of document D_2 (e.g., student file is part of students registry)

Another feature of the information view is that it supports composite information and composite documents. The structuring of information and documents is done via Part Of relationships, allowing analysts to build a hierarchy of information entities or of documents. For instance, this allows representing that the information entities health status and personal information are part of the information medical history info, while information entities present illness and allergies are part of the information medical history of Patient. The same applies to documents. In Figure 4.4, document report is part of document health record.

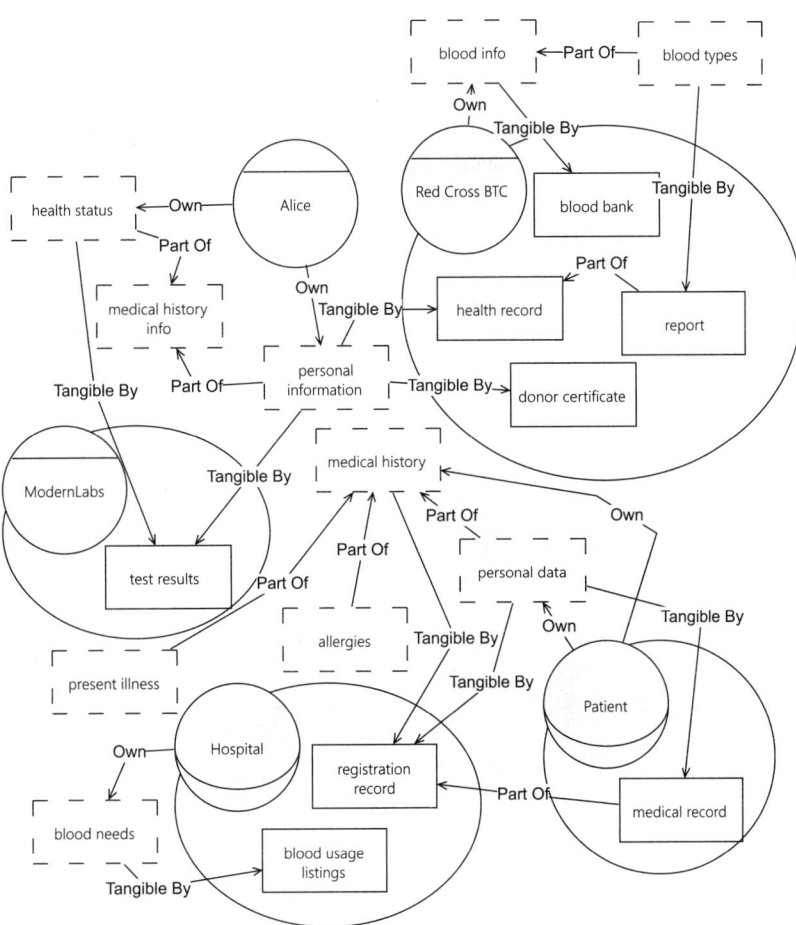

Figure 4.4
Partial STS-ml information view of the healthcare scenario.

4.4   Authorization view

This view supports modeling the authorizations actors grant to others over the informational entities that are modeled in the information view. In STS-ml, an authorization is a directed relationship between two actors, where one actor (*authorizer*) grants or prohibits certain rights to another actor (*authorizee*) on the usage of some information. Authorizations can be defined along four orthogonal dimensions: *allowed/prohibited operations*, *information entities*, *scope of authorization*, and *transferability of the permissions* (see Table 4.8). A partial authorization view for the healthcare scenario is shown in Figure 4.5.

Table 4.8

Authorization view: social relationships.

Graphical notation	Syntax and description
	authorizes($A_1, A_2, \mathcal{I}, \mathcal{G}, \mathcal{OP}, \text{TrAuth}$): actor A_1 authorizes/prohibits actor A_2 to perform operations $\mathcal{OP}$ ($\{\text{R}, \text{M}, \text{P}, \text{T}\} \cup \{\overline{\text{R}}, \overline{\text{M}}, \overline{\text{P}}, \overline{\text{T}}\}$) on the information in $\mathcal{I}$, in the scope of the goals in $\mathcal{G}$, and allows (prohibits) A_2 to transfer the authorization to others if TrAuth is true (false); In the figure, A_1 authorizes A_2 to read information Info 1 (R is checked), but prohibits modifying such information (M is crossed over), in the scope of goal Goal 1 allowing A_2 to transfer such authorization further (continuous arrow line)

For instance, in Figure 4.5, Alice authorizes Red Cross BTC to read information health status, but prohibits the modification of this information, for goal donor approved, granting a transferable authorization. The authorization relationship from Patient to Physician is the most restrictive, for it prohibits all operations and the transferability of the authorization. The Hospital authorizes the Physician to read the information present illness and medical history for goal medical advice given, but does not grant authorization transferability (further granting permissions to other actors).

As introduced already in Chapter 3, the authorization relationship supports a variety of security requirement types. In the following subsection, a description is provided of which supported security requirements from Figure 3.16 are expressed when modeling authorizations.

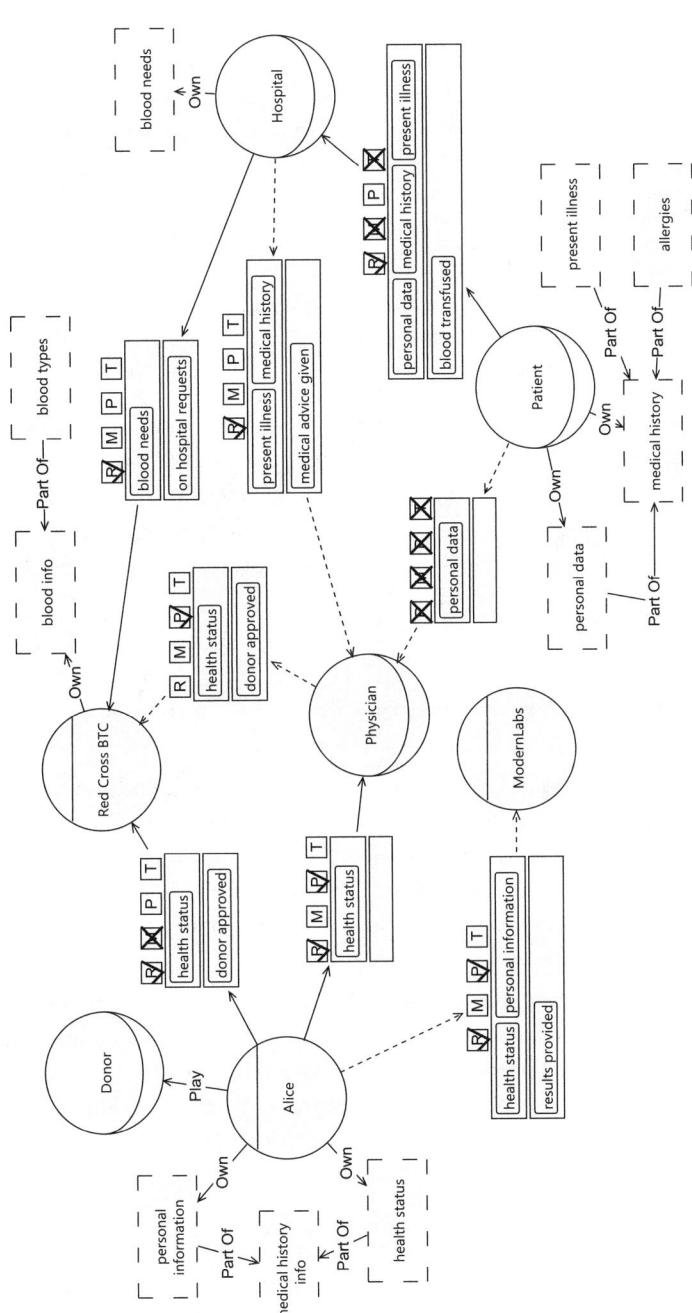

Figure 4.5
Partial STS-ml authorization view of the healthcare scenario.

Implicitly expressed security requirements. Security requirements are expressed through authorizations when certain operations (i.e., read, modify, produce) are prohibited, when the scope is limited to specific goals, or when reauthorization is not granted (further transferring permissions).

Let *Auth* stand for authorize($A_1, A_2, \mathcal{I}, \mathcal{G}, \mathcal{OP}, TrAuth$), where A_1, A_2 are actors, $\mathcal{I}$ is a set of information, $\mathcal{G}$ is a set of goals, $\mathcal{OP}$ is the set of allowed and prohibited operations $\{R, M, P, T\} \cup \{\overline{R}, \overline{M}, \overline{P}, \overline{T}\}$, and *TrAuth* is a boolean value determining if the right to transfer authorizations is granted:

- $\mathcal{G} \neq \emptyset \rightarrow$ *Need-to-know*: the authorizee should not perform any operation (read/modify/produce) on documents that make some information in $\mathcal{I}$ tangible, for any goals not included in $\mathcal{G}$. The authorization from the Patient to the Hospital is an example: personal data, medical history, and present illness can be read only when achieving goal blood transfused (see Figure 4.5).

- $\overline{R} \in \mathcal{OP} \rightarrow$ *Non-reading*: the authorizee should not read documents representing information in $\mathcal{I}$. For instance, the Patient requires this when prohibiting the Physician to read the information personal data in Figure 4.5.

- $\overline{M} \in \mathcal{OP} \rightarrow$ *Non-modification*: the authorizee should not modify documents that include information in $\mathcal{I}$. The authorization from the Patient to the Hospital prohibits the modification operation; thus, no documents containing the information personal data, medical history, and present illness should be modified by the Hospital, irrespective of its currently pursued goals.

- $\overline{P} \in \mathcal{OP} \rightarrow$ *Non-production*: the authorizee should not produce any documents that include information in $\mathcal{I}$. For example, in Figure 4.5, the Patient expresses a non-production requirement on personal data to the Physician, by prohibiting the production operation. That is, the Physician should not produce any new documents containing the information personal data.

- $\overline{T} \in \mathcal{OP} \rightarrow$ *Non-disclosure*: the authorizee should not transmit (disclose) to other actors any document that includes information in $\mathcal{I}$. For instance, the Patient requires this in the authorization over the information personal data, medical history, and present illness allowed to the Hospital. That is, the Hospital should not transmit any document containing any personal data, medical history, or present illness to any other actor.

- $TrAuth =$ `false` $\to$ *Not-reauthorized*: the authorizee should not redistribute the permissions to other actors. If the authorizee receives an authorization that contains only prohibitions, then the requirement non-reauthorized does not apply. An example is the authorization from Alice to ModernLabs, which does not grant a transferable authorization. That means, ModernLabs is itself allowed to read and produce the information health status and personal information, but it should not further pass these rights to other actors.

4.5   Chapter summary

This chapter has presented the three views supported by the STS method to create STS-ml models. The modeling of these three views results in the creation of the overall model for the system-to-be.

The social view represents actors and their interactions; the information view depicts actors as information owners as well as how their information is structured; while the authorization view expresses the permissions and prohibitions actors specify over their proprietary information to other actors in the system.

The information view acts as a bridge between the social and authorization views: in the social, actors manipulate and exchange documents; in the authorization, actors specify their authorizations over information. This distinction and the refinement of information entities supports expressing fine-grained security requirements, in particular over information.

Through its multi-view modeling approach, STS-ml is geared toward modeling non-trivial scenarios, where different perspectives of security have to be accounted for. Chapter 7 shows how this modeling paradigm is effectively supported by specific tooling. Multiple views are also thought to promote collaboration among requirements analysts and security engineers who can create different diagrams of different parts of the system-to-be, instead of exchanging the same diagram.

4.6   Exercises

Review questions

Q4.1. In which view is the information flow captured in STS-ml? Through which primitives?

Q4.2. What is the difference between the social view and the authorization view?

Q4.3. How does the social view differ from the information view?

Q4.4. In which view(s) are stakeholders' assets represented in STS-ml?

Q4.5. In which view(s) are threats represented in STS-ml?

Q4.6. In which view(s) are security requirements captured in STS-ml?

Q4.7. Explain what implicitly expressed security requirements are and how they differ from explicitly expressed ones.

Q4.8. What are the main reasons for having more than one view in a modeling language?

Problems

P4.1. Consider the healthcare scenario introduced in Section 1.5. Suppose that research centers have discovered a new way to remove bacteria from platelet concentrates. The Red Cross BTC plans to provide the research centers an adequate quantity of blood samples to test the efficacy of this method. Conduct the following tasks:

a. Define stakeholders' assets in this extended scenario. How are they represented?

b. Represent information and information owners.

c. What is the information hierarchy (structure)? What is the document hierarchy (structure)?

d. What are realistic permissions and prohibitions that the hospital would impose on others? What about the Red Cross BTC?

e. Can the hospital authorize research centers? What are their rights?

f. What are the permissions the Red Cross BTC has on donors' medical history? Can the Red Cross BTC transfer rights to research centers? What rights?

P4.2. Consider a travel agency service that offers customers the possibility to check various destinations, book flights and hotels and so on. Customers may decide to check the various options offered by the travel agency in booking flights and hotels, or they might do the same by directly checking options online.

a. Pick a destination and represent the various alternatives customers have to organize the trip and stay in the selected destination. For each alternative, identify security issues.

b. Identify the information flows, with a particular focus on security-relevant information.

c. What are the most relevant security issues with respect to the identified information? Explain why.

III FROM STS-ml TO THE STS METHOD

5 Automated Analysis of STS-ml Models

Models are useful abstraction mechanisms to compactly describe functions and desired properties that a system under design should fulfill. This is especially true in requirements engineering, where the analysts often use different types of models to understand what the requirements are for the system-to-be (see Chapter 9 for an overview). In addition to facilitating the understanding of the system and its domain, models can be used for analysis purposes, that is, to answer questions that determine whether certain (un)desired properties hold.

However, requirements models tend to be too large and complex in real-world settings to enable their analysis by hand. Security requirements models are no exception, as demonstrated by our own experiences with early versions of STS-ml [60]. One way to deal with such complexity is to support automated analysis, that is, the use of software tools that are capable of checking whether the (un)desired properties hold.

STS-ml features the execution of a variety of automated analysis techniques over its models, enabling the analyst to:

- Perform *well-formedness* checks: given that goal models tend to become large and complex, well-formedness analysis helps the security requirements engineer to build syntactically correct models, in other words, models that follow the syntax of the STS-ml modeling language.

- Identify conflicts between security requirements as well as possible violations of security requirements. This type of analysis is called *security analysis*; this is intended to identify conflicts arising due to (i) conflicting security requirements that a given participant is expected to satisfy, and (ii) security requirements that a participant cannot satisfy without violating its own objectives.

- Calculate the effect of social threats on stakeholders' assets. The representation of social threats alone would not be enough to know the effect they have on stakeholders' assets. *Threat analysis* helps the analyst by determining the trace of a given threat, that is, the chunk of an STS-ml model that is affected by the occurrence of an event.

The following sections detail why each type of analysis is needed, and how analysis results are used to improve STS-ml models.

5.1   Model well-formedness analysis

The purpose of well-formedness analysis is to verify whether the diagram built by the designer, defined in terms of the concepts and relationships in the three views, follows the syntax of STS-ml. This analysis is integrated in STS-Tool (Chapter 7) to support security requirements engineers interactively while building an STS-ml model.

Some rules are verified on-the-fly (as soon as the modeler adds or removes an element) to ensure correctness. For instance, goal-document intentional relationships (reads, modifies, produces) are allowed only from a goal toward a document to indicate that the operation (reading, modifying, producing) is performed on the document while fulfilling the goal. These rules are a straightforward application of the syntax that is presented in Chapter 3.

Other rules are computationally too expensive for on-the-fly verification, or their continuous analysis would limit the flexibility of the modeling activities by over-restricting the models that can be created. Thus, some analysis techniques about well-formedness are performed upon explicit user request. In the following section, these types of checks are described, highlighting their relationship with other types of checks, and the actions that the security requirements engineer should perform when the analysis returns an error.

5.1.1   Empty diagram

This check verifies whether or not the given diagram is empty. If the diagram is empty, that is, no elements are drawn, then no other well-formedness checks are possible. If the diagram is not empty, the analysis returns: "No errors found." When the model is empty, the modeler simply has to add at least one element to the diagram.

5.1.2   Goal single decomposition

This check verifies the consistency of goal decompositions. Following the semantics of STS-ml, a decomposed goal should lead to two or more subgoals. Therefore, this analysis verifies whether there are cases of decompositions to a single subgoal and returns an *error* should that be the case. Recall from Section 3.2.3 that goal decompositions define either how a goal is fulfilled via subgoals (and-decomposition), or alternative ways to fulfill the goal (or-decomposition). Therefore, the presence of just one subgoal would either not define the steps to fulfill the decomposed goal (in case of and-decomposition), or present no alternatives for fulfilling the goal (in case of or-decomposition).

5.1.3   Delegation child cycle

This check verifies the consistency of goal delegations, so that no cycles or loops are identified as a result of the delegatee decomposing the delegatum (delegated goal) and redelegating back one of the subgoals. Delegation cycles would result in an STS-ml model that does not clarify which actor is responsible for fulfilling the delegated goal. Notice that this situation could actually occur when the security requirements engineer emphasizes that a delegatee might not accept the full delegation of the goal and gives back part of it. Thus, this verification returns a *warning* instead of an *error*, even when a cycle is found. The warning informs the security requirements engineers, but they are left to decide whether to address the warning or to ignore it.

5.1.4   Documents part-of cycle

This verification is about whether there is a cycle of part-of relationships starting from and ending at the same document. If a case like this is verified, a warning is returned enumerating the documents that form the cycle. Verifying the existence of document part-of cycles is important to avoid ambiguities in reasoning over violations of security requirements, more specifically confidentiality requirements, such as non-reading and non-disclosure. Indeed, when document D_1 is part of D_2, and D_2 is part of D_1, one would conclude that $D_1 = D_2$, thereby expecting them to represent the same information, be subject to the same security needs, and so forth.

5.1.5   Information part-of cycle

This check verifies whether there is a cycle of part-of relationships starting from and ending at a given informational entity. If a case like this is verified, a warning is returned enumerating the documents that form the cycle. Cycles of part-of relationships over information entities could result in ambiguities over information ownership, just as in the case of documents part-of.

5.1.6   Information without ownership

This check verifies that all information entities have an owner. If there are cases of information without any ownership relationships (direct or inferred through part-of relationships) from any actor in the diagram, the well-formedness analysis will return a warning. The reason for this is that information owners are the ones originally authorized to express security needs about information, unless such authorization is transferred. In many cases, this verification informs the security requirements engineers about any missing ownership links to ensure that these were not forgotten. But if the information is indeed not owned by any actor, then the model requires no changes. Hence, this check falls under the category of warnings, not errors.

5.1.7   Authorizations validity

This check verifies that all authorization relationships between two given actors are valid. An authorization relationship represents permissions and prohibitions that an actor specifies to another on some information, to/not to perform some allowed/prohibited operations, possibly limited to a goal scope and with authority for further redelegation (see Section 3.3). For an authorization relationship to be valid, the first two attributes, that is, allowed/prohibited operations and information entities, should be specified. When no information is specified, the authorization is not expressing the information for which authorization is granted or prohibited. Similarly, if nothing is specified over operations, then it is not clear what the authorizee actor can and should not do with the specified information. Thus, the findings for this check fall under the category of errors.

5.1.8   Duplicate authorizations

This check verifies that there are no duplicate authorization relationships. Two cases are addressed by this check:

1. Two identical authorization relationships, that is, between the same actors, in the same direction; for the same set of information entities, operations, and goals; and having the same value of transferability
2. Authorization relationships between the same actors, in the same direction, in which one grants permissions that are a subset of the other authorization's relationship

This check is inspired by the need to keep the models clean and concise, and it aims to identify authorization relationships that could be merged into a single one. Given the cognitive scalability limits of modeling languages, whenever possible, one should avoid redundancies. This check leads to warnings.

5.2 Requirements conflict analysis: security analysis

Security analysis verifies whether or not the drawn diagram allows the satisfaction of the specified security requirements. For all security needs expressed by the stakeholders, it checks in the model whether there is any possibility for the derived security requirements to be violated. This is done at two levels:

1. Identifying possible conflicts among security requirements, that is, two or more requirements that cannot all be fulfilled by the same system
2. Identifying conflicts between the actors' respective business policies— i.e., how they achieve their goals—and the security requirements imposed on them by other actors

In the following subsections, the details for all the checks that belong in security analysis are provided.

5.2.1 Conflicting authorizations

This check verifies if the stakeholders have expressed conflicting authorizations. This is likely to happen because STS-ml models allow for multiple authorizations (either permissions or prohibitions) over the same information.

For instance, in the healthcare scenario, there is an authorization from the Patient to the Hospital, in which the Patient allows the Hospital to read (checkmark symbol over R, Figure 4.5) the information personal data, medical history, and present illness but prohibits the modification and transmission (X symbol over M and T) of the given information, and specifies nothing on production. The authorization is limited to the scope of goal blood transfused. If another authorization existed for the Hospital for the same information with no restrictions on the goal scope, then an authorization conflict would exist. Similarly, if more authorizations for the Hospital existed, prohibiting the granted operation (in this case R) or granting the right to perform any of the prohibited operations (here, M and T) over any of the specified information entities (personal data, medical history, or present illness) or parts of these information entities, in the scope of the same goal (i.e., including its subgoals and descendants in the goal tree), then an authorization conflict would arise. The check on conflicting authorizations enables the verification of these scenarios.

An authorization conflict occurs if both authorizations apply to the same information and either of the following applies:

1. One authorization restricts the permission to a goal scope, while the other does not; that is, one implies a need-to-know security requirement, while the other grants rights for any purpose and therefore does not impose such a requirement.

2. The scopes are intersecting and contradictory permissions are granted on operations (read, modify, produce, transmit), or authority to transfer (transferability of authorization).

Figure 5.1 illustrates this analysis detecting two authorizations for Physician on personal data: (i) that from the Hospital, which grants the right to read (R) and specifies nothing on modification (M), production (P), or transmission (T) of the information present illness and medical history (personal data is part of medical history); (ii) that from the Patient, which prohibits all operations R, M, P, and T on the information personal data. The first authorization specifies a need-to-know security requirement in the scope of goal medical advice given, while the second authorization does not specify anything, since it imposes only prohibitions over the information personal data. Thus, following the authorizations conflict check, the two authorizations have intersecting scopes (the scope of the first authorization) and they are both specified over the information personal data. Confronting the granted and prohibited operations from each, the analyst can deduce that they are conflicting with respect to the read (R) operation for information personal data.

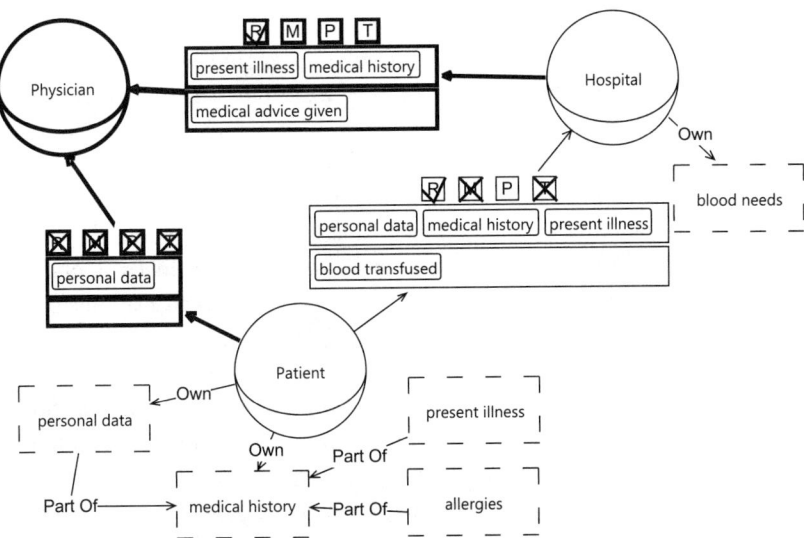

Figure 5.1
Authorizations conflict, highlighted with thicker black lines.

The identification of authorization conflicts allows security requirements engineers to fix them by interacting with stakeholders to identify
resolutions.

5.2.2   Conflicts between business policies and security requirements

Each actor model defines a specific actor's business policy, that is, the
actor's goals and its alternative strategies to achieve these goals (see
Section 3.2.3). Given an STS-ml model, it is possible to verify if any
security requirement is violated by the business policy of any actor in
the model. For instance, the requirement non-reading(Alice, Red Cross BTC,
{personal information}) would conflict with a business policy for Red Cross
BTC that includes the goal-document relationship reads(Red Cross BTC,
donor approved, personal information).

An actor's *business policy* defines alternative strategies for an actor to
achieve its root goals. It is a sub-model of the social view that includes
all the goals and documents in the scope of that actor in the social view,
the relationships (and-/or-decomposition, reads, modifies, and produces)

among those goals and documents, as well as goal delegations and document transmissions that start from that actor.

In the healthcare running example, the business policy of Hospital includes goals health service provided, which is and-decomposed into subgoals health records maintained and blood transfused. The first is further and-decomposed into subgoals patient registered, which produces document registration record, and medical history obtained; while the second (goal blood transfused) reads document blood bank, produces document blood usage listings, and is or-decomposed into transfusion performed via specialist and transfusion performed in surgery (see Figure 5.2). The or-subgoals denote alternative strategies: the actor can choose either of them to achieve the upper level goal. Hence, Hospital may choose to perform transfusion procedures either with the help of a specialist or during surgery.

Given an actor's business policy, alternative strategies are introduced by (i) choosing one subgoal in an or-decomposition and (ii) deciding whether to pursue root goals that are delegated from other actors. In Figure 5.2, the business policy for Hospital includes one root goal and one delegated goal patient registered, which is not a root goal. Therefore, one strategy for the Hospital involves achieving all and-decomposed goals, excluding the sub-tree rooted by goal blood transfused, which is or-decomposed. Thus, one subgoal is retained in the strategy (e.g., transfusion performed via specialist), while other goals are pruned (in this case, transfusion performed in surgery). The reads relationship to document blood bank is retained, as well as the document itself. The same applies to the produces relationship and the document blood usage listings. An alternative strategy could, however, involve not performing the transfusion via specialist but in surgery. This strategy would differ from the previous only in the choice of the child goal of blood transfused.

Similarly, ModernLabs has to achieve tests taken and has in its strategy all goals infect disease tested, drug test performed, and results provided, for these are and-subgoals of its root goal. However ModernLabs can choose to achieve results provided through results retrieved in person or results sent via courier. Figure 5.3 shows these alternatives in ModernLabs' actor model.

The identification of actors' strategies allows verifying whether these strategies conflict with the security requirements that are imposed on the actors. For a given actor, a conflict between its business policy and security requirements occurs in two cases:

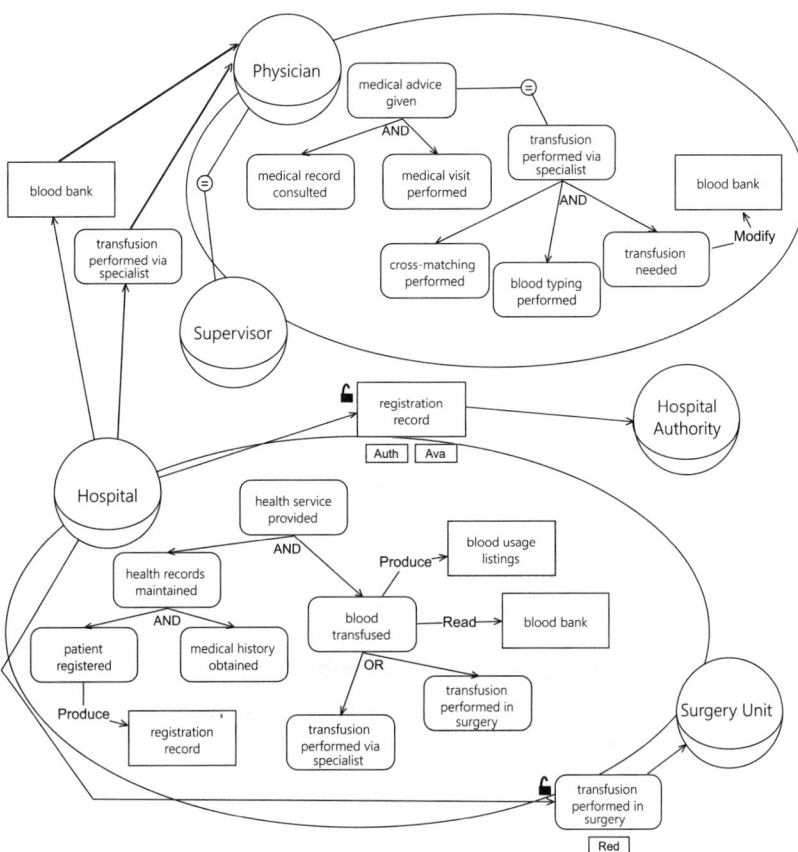

Figure 5.2
Partial business policy of roles Hospital and Physician.

1. The actor's strategy contains one or more relationships (operations performed by the actor as denoted by its business policy) that are prohibited by a security requirement requested by another actor.
2. The actor's strategy does not contain any relationships (operations performed by the actor to pursue its goals) stipulated by some requirement requested by another actor.

The requirements for each security requirement category (Figure 3.16) are discussed below, including those requirements that necessitate verification of runtime.

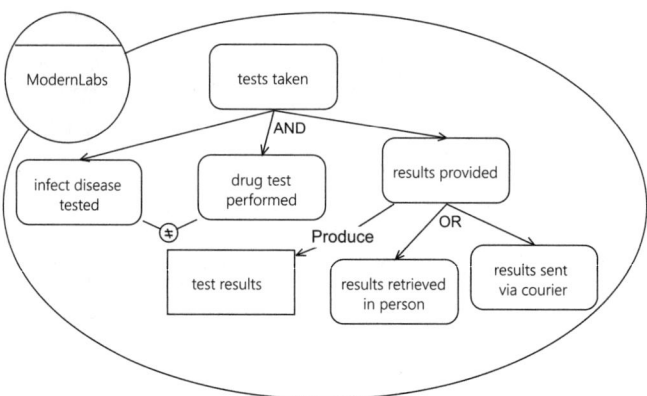

Figure 5.3
Partial business policy of agent ModernLabs.

5.2.2.1   Accountability requirements

Non-repudiation requirements over goal delegations and document transmissions cannot be verified at design time. Indeed, they require the ability to check the occurrence of runtime actions; for example, a requirement for non-repudiation of acceptance applied to a goal delegation requires checking that a proof of fulfillment was sent and received.

Non-redelegation is verified if there is no delegation relationship having the goal G or its subgoals as delegatum (delegated goal), from the delegatee A_2 to other actors. Therefore, to check for any violations of non-redelegation, the analysis searches for redelegations of the delegatum or any of its subgoals. In Figure 5.4, Alice expresses a non-redelegation security requirement over the delegation of goal tests taken to ModernLabs, and the latter delegates goal drug test performed to Drug Tests Inc, a goal that is the and-subgoal of tests taken. This scenario constitutes a violation of non-redelegation.

Role-based separation and combination of duties require all actors *not to* and *to* play two roles through plays relationships:

- A role-based SoD violation occurs whenever a single actor plays both roles among which an SoD constraint is expressed. Recall that role-based SoD requires that no agent can play both roles if it plays one of them.

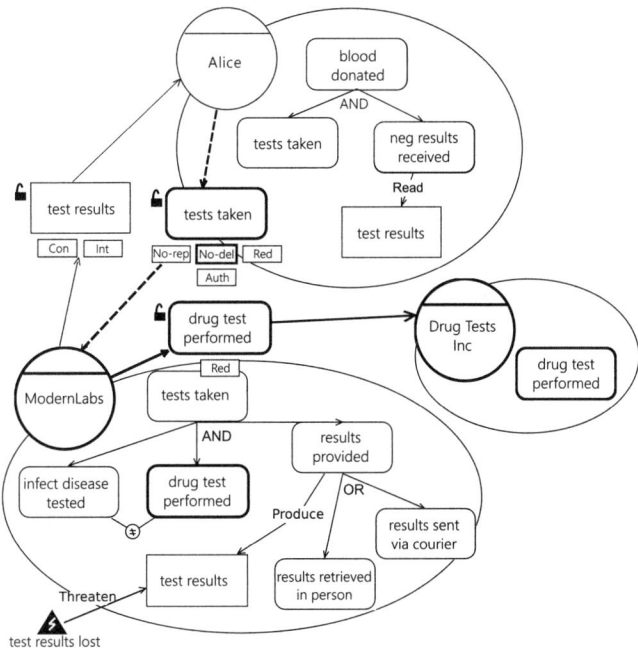

Figure 5.4
Accountability requirements: example of non-redelegation violation.

- A role-based CoD violation occurs when there is no agent that plays both roles among which a CoD constraint is expressed. Recall that role-based CoD requires the same agent to play both roles if it plays one of them.

Goal-based separation of duties is verified if no actor pursues both goals or subgoals in the requirement. Thus, a violation of this requirement is detected whenever a single actor may perform both goals for which a goal-based SoD security requirement is expressed. Goal-based SoD requires that no actor perform both goals among which SoD is specified. To perform this verification, the analysis checks whether the final performer of the given goals is the same actor.

Goal-based combination of duties is verified in a similar way: a violation occurs whenever a single actor may not perform both goals for which a goal-based CoD is expressed. Goal-based CoD requires the same actor to perform both goals among which CoD is specified. The analysis verifies whether the final performer of the given goals is not the same actor.

5.2.2.2   Reliability requirements

Redundancy requirements can be partially checked at design time. The existence of redundant alternatives can be verified, but an actor's strategy does not tell how alternatives are temporally interleaved, that is, whether they provide true redundancy (when the alternatives are both made available concurrently) or fallback redundancy (when an alternative is activated after the principal available fails). As a result, true and fallback redundancy are checked in the same way at design time.

Single actor redundancy is fulfilled if the delegatee has at least two disjoint alternatives (via or-decompositions) for the specified goal. Multiactor redundancy requires that at least one alternative involve another actor different from the delegator and the delegatee. Specifically, the security analysis verifies a redundancy violation if one of the following occurs:

1. The delegatee actor does not or-decompose the delegated goal; both types of redundancy are violated.
2. The delegatee actor decomposes the goal into or-subgoals and delegates one to another actor when single actor redundancy has been specified; this type of redundancy is violated.
3. The delegatee actor decomposes the goal into or-subgoals but does not delegate any of the subgoals to another actor when multi-actor redundancy has been specified; this type of redundancy is violated.

Trustworthiness requirements cannot be verified at design time, since they require the delegatee to provide at runtime a proof of trustworthiness, such as proof issued by a certification authority.

5.2.2.3   Authenticity requirements

Delegator/sender authentication is typically implemented in electronic commerce websites, wherein a certification authority guarantees the authenticity of the merchant's website. Therefore, the verification of such a requirement involves actions and mechanisms that can be verified only at runtime, when the system is deployed.

Delegatee/receiver authentication is a situation we encounter every day when browsing the web and using our credentials (username/password) to access web information such as our email. The fulfillment of this type of requirement cannot be verified at design time.

5.2.2.4 Availability requirements

Verifying availability requirements (both goal and document availability) calls for measuring the availability level. Notice that goal availability is highly related to the notion of service availability, where a provider specifies an uptime level for the service (look back at Table 3.2). In service-oriented settings, availability levels often become part of service-level agreements between providers and consumers. These requirements cannot be checked at design time, as they require runtime monitoring.

5.2.2.5 Integrity requirements

Non-modification requires that the authorizee's strategy include no modifies relationships on documents that make tangible part of the information entities specified in the authorization. Non-modification expresses the need for the information not to be changed (modified); that is, authority to modify the information is explicitly prohibited, or there are no authorizations to grant the authority to modify. This violation occurs whenever an actor modifies information without having the right to modify it. To verify if there could be any violations of non-modification, the analysis checks if the authorizee (or an actor that is not authorized by authorized party) modifies the given information. For this, it searches for modifies relationships from any goal of this actor to any document representing the given information. The analysis identifies several non-modification violations. One is that of Physician, who modifies information blood info without having the right to modify (see Figure 5.1). Modification takes place because of the modify relationship from the Physician's goal transfusion needed toward document blood bank, which represents blood info (see Figure 4.4).

Integrity of transmission requirements prescribe that the integrity of document transmission be preserved, and this can only be verified at runtime.

5.2.2.6   Confidentiality requirements

These are expressed through different types of authorizations and prescribe the set of relationships that should not be present in the authorizee's strategy.

Need-to-know is verified by the absence of reads, modifies, or produces relationships on documents that make tangible some information in the set of authorized information entities for some goal that is not in the scope of authorization (or any subgoals of the goal scope).

Requirements non-reading and non-production are verified if the authorizee's strategy includes no read or produces relationships on documents that make tangible part of the authorized information entities. The analysis identifies several violations of non-reading in the running scenario. For instance, the Hospital violates this requirement because it reads information blood info (there is a read relationship from Hospital's goal blood transfused toward document blood bank representing the information blood info in Figure 5.2), while Hospital does not have the authority to read (in Figure 5.1, the Hospital has no authorization granting the right to read that information). The analysis has not found any violations of non-production.

Non-disclosure does a similar check but looking at document transmissions, that is, transmits relationships. In the healthcare scenario, in Figure 5.2, Hospital violates non-disclosure by transmitting document blood bank (representing information blood info) to Physician without having the right to transmit such information (there is no authorization relationship toward Hospital).

Non-reauthorization is fulfilled if there is no authorizes relationship from the authorizee to others on any operation.

Confidentiality of transmission can only be verified at runtime because its verification involves mechanisms for checking that the confidentiality of the transmitted document is preserved.

5.3   Threat analysis

The purpose of this analysis is to assess the effect of events threatening actors' assets. As introduced in Section 3.4, events may threaten actors' supporting assets (subgoals and documents) to exploit their primary assets (root goals and information). Threat analysis answers the question: *how does an event threatening any actor's supporting asset affect the rest of an STS-ml model?*

Threat analysis focuses on events threatening goals or documents; as such, it propagates the effects over goal trees and goal-document relationships within an actor model (reads, modifies, produces), as well as social relationships involving goals and documents (delegates, transmits). Therefore, starting from the specified threats, the technique identifies their effects over the rest of the diagram. The analysis starts with the known events and propagates their effect over goal trees, documents, and social relationships. The newly discovered elements are treated as threatened elements. The analysis ends when no new elements are found. The propagation rules are the following:

1. If an event threatens a goal and (a) the goal is an and-subgoal of another goal, then the latter goal is considered threatened; (b) the goal is delegated, then the threat is propagated to the goal of the delegator; (c) the goal produces a document, then the threat is propagated to the document too.
2. If an event threatens a document and (a) the document needs to be used or modified by some goal, then the goal is considered threatened too; (b) the document is transferred to some other actor, then the sender's document is threatened too; (c) the document is composed of other documents, then the threat is propagated to those parts of the document.

In the healthcare scenario, the event test results lost threatens Modern-Labs' document test results (see Figure 5.5). By executing threat analysis, the analyst determines the effects of this event on the rest of the actors. For instance, the event test results lost threatens Alice, one of the actors interacting with ModernLabs. The latter transmits the document test results to Alice, and since this document is threatened, then Alice's goal neg results received is threatened too, since it needs to read test results. Similarly, goal blood donated is threatened since its and-subgoal (neg results received) is threatened.

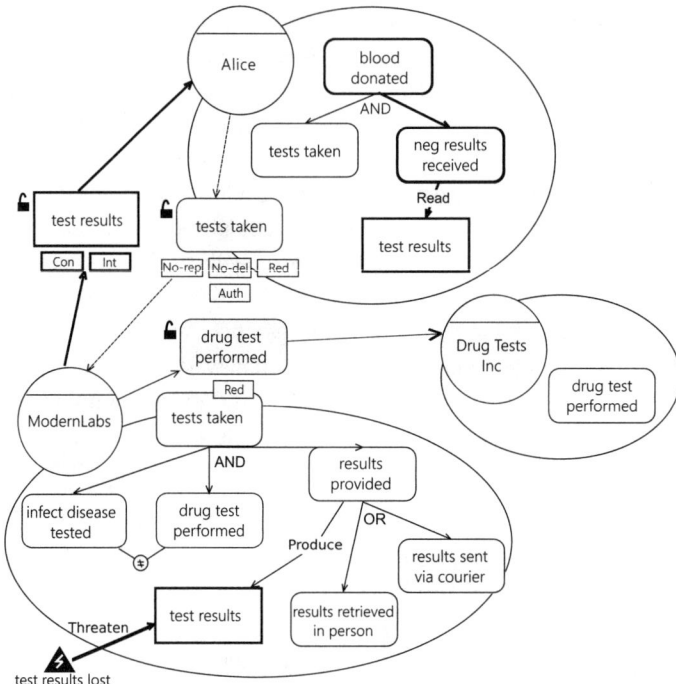

Figure 5.5
Threat propagation example.

5.4   Improving a model through analysis

The analysis techniques in the previous sections serve the purpose of identifying problems in the created STS-ml model. The security requirements engineer can use the obtained results to improve the models in an iterative and incremental fashion, starting from an early and partial version of the STS-ml model.

5.4.1   Well-formedness analysis

As a rule of thumb, well-formedness analysis (Section 5.1) should be executed first, for the other two analysis techniques verify security aspects on a well formed model.

In the healthcare scenario, the execution of the well-formedness analysis identifies warnings and errors. For instance, in Figure 4.1 there cannot be a delegation from Physician back to Hospital of goals transfusion performed via specialist (delegation child cycle, Section 5.1.3). Similarly, in Figure 4.4 there cannot be a part-of relationship from information medical history to information personal data (information part-of cycle, Section 5.1.5). The authorization from Physician to Red Cross BTC would not be well-formed, if the right to produce were not specified (authorizations validity, Section 5.1.7).

5.4.2 Security analysis

5.4.2.1 Authorization conflicts

The identification of authorization conflicts helps detect inconsistencies among security requirements, which might have consequences on actors' behavior. Let us consider an actor A. If another actor A' grants A an operation, and another actor A'' prohibits him performing the same operation, what is actor A supposed to do? Performing the operation would result in violating A''''s imposition. Not performing the operation may conflict with A's own business policy. Therefore, it is crucial that such conflict be identified and then resolved.

Resolving the conflict requires one of the actors to change its specified authorizations. In the healthcare scenario, an authorization conflict involves actor Physician for information personal data on the right to read, because the Patient has prohibited this operation, while the Hospital has granted (allowed) this operation. To resolve this conflict, either the Patient decides to grant the Physician the right to read the information personal data, or the Hospital specifies a non-reading security requirement too, prohibiting the right to read. This decision is made by negotiating with the involved parties to reach a consensus that best fits their needs.

5.4.2.2 Conflicts between business policies and security requirements

The identification of these conflicts helps avoid violation of security requirements due to inconsistencies with an actor's business policy. In the healthcare scenario, several such violations exist. For instance, a violation of non-modification exists from the Physician on information blood types. Resolving this inconsistency requires negotiating the rights the Physician should have. If modifying the blood types is required for the Physician to perform his duties, then this right should be granted to him (consequently, the non-modification security requirement is no longer specified). Otherwise, if information blood types should remain unchanged and taken as is, then the Physician should change his business policy (the modifies relationship should be omitted).

5.4.3   Threat analysis

Identifying how events affect stakeholders' assets contributes to mitigating the impact of these threats. For instance, in the healthcare scenario, an analyst can determine the effects of the event test results lost on ModernLabs and Alice. One way to deal with this threat is to have alternative ways to provide test results for approving a prospective Donor's blood. In the current situation, ModernLabs transmits the test results to Alice to have her blood approved so she can donate blood (blood donated). An alternative strategy could consider that another copy of test results might be kept by ModernLabs.

5.5   Chapter summary

Modeling languages are useful means to represent knowledge, but as they grow in size, the chance of becoming inconsistent increases significantly. STS-ml models are no exception to the rule. The detection and handling of conflicts between requirements is a hard task [17] due to the increasing size and complexity of models. This is true in STS-ml too: the gained expressiveness and the richer set of supported security requirements come with a price—conflict identification cannot be performed by merely looking at the models.

Handling these conflicts is crucial to avoid going to the design phase with an inconsistent specification of security requirements that no system can possibly satisfy. Automated reasoning techniques help in identifying inconsistencies and errors in the models.

In this chapter, a number of analysis techniques for STS-ml have been presented, spanning the identification of security requirements conflicts, the discovery of conflicts among actors' business policies and the security requirements they have to comply with, and the determination of the threat trace for events threatening actors' assets. These activities are important to avoid developing a system that violates some security requirements. Knowledge about threats and their effects on stakeholders' assets being spread across the socio-technical system is crucial to anticipate and ideally avoid potential problems in a running system. These analyses are all integrated in STS-Tool (see Chapter 7) to support automated analysis of STS-ml models.

5.6   Exercises

Review questions

Q5.1. When is an STS-ml model well formed?

Q5.2. Why is it important to verify well-formedness?

Q5.3. What is covered in security analysis? Discuss the categories, identifying their differences.

Q5.4. How is the impact of threats calculated over STS-ml models?

Q5.5. Why are automated reasoning techniques useful in the process of creating STS-ml models?

Q5.6. Provide three examples of authorization conflicts in the context of mobile banking.

Q5.7. Explain the notion of a conflict between business policies and security requirements. Provide one example.

Problems

P5.1. Consider the STS-ml model of the healthcare scenario that you created as an answer to problem P4.1. Answer the following questions:

a. Is the model well formed? Why?
b. Is there any case of unauthorized access to donors' sensitive information?
c. Is there any case of unauthorized disclosure of donors' medical records?
d. Are there any unauthorized delegations of rights?

      e. What happens if information on donors' health conditions cannot be confirmed?

P5.2. Consider a travel agency service that offers customers the possibility to check various destinations, book flights and hotels. Use the STS-ml model you created to answer problem P4.2, or create the model now. Answer the following questions:

      a. How does the well-formedness analysis help you improve the model? Conduct multiple iterations until the issues are fixed.

      b. Identify three violations of confidentiality requirements. Can these violations be fixed? How?

      c. Are there any violations of availability requirements? What about reliability requirements? Why do these violations occur?

6   The Socio-Technical Security Method

This chapter presents the Socio-Technical Security (STS) method for security requirements engineering. This method uses the STS-ml language as its main artifact: the analysts construct socio-technical models of the system that, eventually, lead to a security requirements specification for the system under design.

After presenting a high-level overview of the method in Section 6.1, and showing how STS can be embedded within broader software and systems engineering methods (Section 6.2), the chapter continues with a description of the process and the involved roles (Section 6.3), and with a detailed elaboration on each of its five major phases (Section 6.4 to Section 6.8).

6.1   Method overview

A high-level overview of the process that the STS method follows is presented in Figure 6.1, showing how the whole security requirements engineering phase is covered, the principal roles conducting and participating in each phase (the icons associated with each of the phases), and the iterative and incremental nature of the method (see the spiral shape in the background).

Unlike other methods for SRE, which result in the security specification for a software system, the STS method produces the *security specification for the overall socio-technical system*. This specification is the result of continuous iterations of the five major sub-phases: elicitation, modeling, analysis, specification of security requirements, and their validation.

In line with goal-oriented requirements engineering, the method starts with *elicitation* activities to determine stakeholders' objectives and their security needs. This activity is conducted by the security requirements engineer with the help of stakeholders, through discussions and interviews to identify their security needs. Additionally, further needs are acquired by studying organizational regulatory rules, laws, or any other documents providing related information. In STS, no specific technique to obtain such details is proposed, although doing so is essential to the SRE process.

Being reliant on the STS-ml language, the STS method guides requirements analysts and security engineers (represented in Figure 6.1 by the

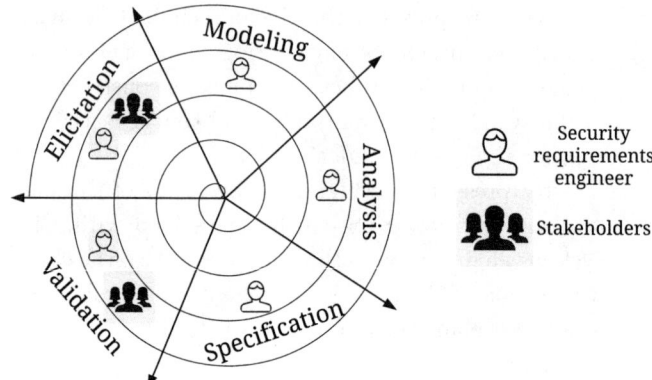

Figure 6.1
The STS method: overview.

security requirements engineer) through an iterative process of *modeling* and *analysis* of the socio-technical system, which leads to the *specification* of a version (list) of the security requirements. These three phases constitute the core of the STS method and are thoroughly analyzed in the rest of this chapter.

The models are refined on the basis of the feedback provided by the analysis techniques and the discussions with stakeholders to *validate* the specification, which could eventually lead to the *elicitation* of additional or revised requirements. Further iterations take place until convergence is obtained, that is, when the security requirements specification is validated by the stakeholders, and no new requirements are elicited thereafter.

In its early phases, the method focuses mainly on the *"why"* and *"what"* questions, that is, why security is important to stakeholders, and what security needs stakeholders have. The STS method provides directions to answer the *"how"* question, that is, how the security needs can be satisfied. However, a complete answer to the *"how"* question requires mapping the STS specification to lower-level languages for business process modeling (e.g., BPMN [44]) and policy specification (e.g., XACML [43], SecPAL [3]). While business process modeling languages are useful to specify the procedural aspects of the security policies among the participants in the socio-technical system, policy specifications can be implemented in a security infrastructure that helps enforce the specification derived from STS.

6.2   STS as part of software/systems engineering methods

The STS method can be included in the context of broader methods, such as those for system engineering and for software engineering. Because of the evolving nature of requirements, and the STS method's iterative and incremental process, it is optimally employed within agile methods that support quick response to changes. In particular, the separation of concerns principle followed for modeling activities (see details below) allows the participation of multiple requirements analysts and security engineers to model various components or parts of a socio-technical system.

However, given the complexity of socio-technical systems and their considerable size (reflecting the number of participating actors and their interactions), resulting models tend to be of a considerable size too. This opens up issues and challenges for the integration of the models, which the supporting software tooling accompanying the method should satisfy.

Broadly, the STS method is appropriate for open and larg-scale systems. As argued in Chapter 1, STS in not intended for any type of software, but rather for socio-technical systems. For instance, an operating system or a compiler are not examples of socio-technical systems. The challenges and issues that come with the STS method are justifiable in the light of complex socio-technical systems, but they are not appropriate for artifacts such as a compiler. Their security specification can be handled by traditional software development techniques, and their deployment is typically handled with technical mechanisms. Instead, STS is intended to shed light on the interrelationships among the various subsystems, be they software, hardware, humans, or organizations, in order to identify security discrepancies and potential breaches arising as a result of social interactions.

6.3   Process and roles

The overall detailed process followed by the method is illustrated in Figure 6.2 using the BPMN 2.0 notation [44]. The iterative and incremental nature of the STS method has already been depicted in Figure 6.1; in the detailed process in Figure 6.2, the focus is on the activities of modeling, analysis, and specification, and only essential feedback loops (referring to iterations of the core STS phases) are represented to keep the model readable.

The BPMN diagram outlines five major phases—*social modeling, information modeling, authorization modeling, automated analysis,* and *specification*; the activities within these phases; the roles that execute the activities—*security engineer* and *requirements analyst*; the artifacts— *social view, information view, authorization view, analysis results,* and *security requirements document*; and two external activities (dashed border) that provide inputs to the core process—*elicitation* and *risk analysis*.

Modeling activities are supported by STS-Tool, which ensures interview consistency (details in Chapter 7). The changes in one view have effects on other views. The different represented stakeholders may be first captured in one view, such as the social view, and maintained throughout the three views, unless the requirements engineer decides to hide a stakeholder in another view, such as the information view, should that stakeholder not have any informational assets.

The process supported by the STS method provides guidelines for the requirements analysts and security engineers to follow when specifying secure socio-technical systems, especially for novices with the STS-ml language. However, the presented steps are not prescriptive. For instance, the requirements analyst may decide to represent stakeholders' informational assets before modeling social interactions among them.

Modeling is succeeded by analysis, which is automated by STS-Tool, and aims to identify inconsistencies and conflicts in the created views. The execution of automated analysis requires some modeling to have taken place, but does not necessitate a complete-finished model. The requirements analyst and the security engineer use the results of the automated reasoning for the stepwise improvement of the STS-ml model.

The process continues till the security requirements engineer considers all important security issues covered when modeling the socio-technical system at hand. The termination criteria is established by addressing the following questions:

- *Did I capture all important stakeholders?*

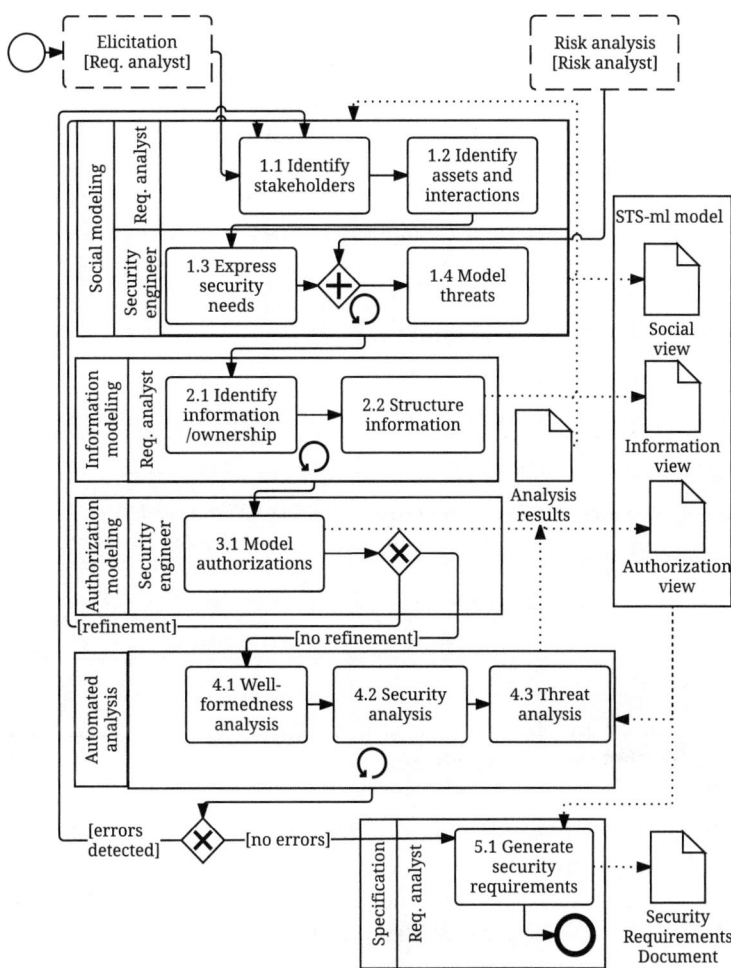

Figure 6.2
The STS method: process.

- *Did I capture all important interactions?*
- *Did I model all assets?*
- *Did I express all security needs?*

The process ends with the derivation of a security requirements specification for the socio-technical system under consideration. At this point, *validation* from Figure 6.1 follows: if the stakeholders agree, the STS method terminates; otherwise, a further iteration of the core process in Figure 6.2 is performed.

Table 6.1 summarizes the three roles involved in the STS method and describes their main responsibilities. Note that the competencies related to the roles *requirements analyst* and *security engineer* might be encapsulated in a single role, that of a *security requirements engineer* (as in Figure 6.1), who possesses the expertise of both. Also note that the risk analyst has a marginal role; its function does not depend on the direct use of STS-ml.

Table 6.1

Roles participating in the STS method and their responsibilities.

Role	Responsibilities
Requirements analyst	Identifying stakeholders, their assets and interactions, their objectives; creating STS-ml models
Security engineer	Investigating the stakeholders' security needs; communicating needs to the requirements analysts; communicating assets and needs to the risk analyst
Risk analyst	Identifying threats; prioritizing threats; communicating relevant threats to the security engineer

6.4   Phase 1: social modeling

Security is mainly concerned with the protection of stakeholders' assets. Aligned with this observation, the STS method starts with a study of the context—led by the requirements analyst—which encompasses identifying the stakeholders in the socio-technical system (Activity 1.1 in Figure 6.2) and their assets, and the interactions among actors (Activity 1.2), based on the inputs of the *elicitation* phase (see Figure 6.1).

As explained in Chapter 3, Section 3.2.2, STS-ml supports two types of assets: *informational*, information that the actors wish to protect; and *intentional*, objectives that they want to achieve. In the social view, information is modeled in terms of the documents that represent it, which the stakeholders possess, manipulate, and exchange. The protection of documents as assets becomes evident when considering the confidential information they contain. In the motivating scenario, an informational asset is the Patient's document medical record, while an intentional asset for Red Cross BTC is goal blood collected.

After defining the context, the security engineer leads Activity 1.3: eliciting and expressing the stakeholders' security needs regarding their interactions. For instance, Alice may want to achieve tests taken only in a specialized laboratory, such as ModernLabs, before filing an application as a Donor, and she may require tests to be examined in that specific laboratory.

Activity 1.4 supports considering possible attackers that could exploit vulnerabilities of the socio-technical system. Potential threats are identified; for instance, test results lost might threaten the possibility of delivering (and thus, receiving) test results on time. This threat affects ModernLabs in providing a timely service to its Patients, including specific ones such as Alice.

Note that threats are social and organizational; they do not necessarily exploit technical vulnerabilities of a software system. Let us reconsider the threat test results lost. This threat might occur as a result of a staff member forgetting test results at a printer; it is not necessarily the result of someone intruding in the system and stealing the results.

The social modeling phase is conducted iteratively to refine the initial model. The outcome of this phase is the *social view*, which was extensively described in Section 4.2.

6.5   Phase 2: information modeling

STS-ml distinguishes between *primary* and *supporting assets*. In Phase 1, the analysts look at the informational assets in terms of concrete exchanges of documents. These documents are supporting assets, and their relevance from a security standpoint is due to the information (primary asset) they represent.

The objective of information modeling is twofold. First, the requirements analyst identifies the information in the considered domain and determines which stakeholders own this information (Activity 2.1). For instance, medical history is information owned by the Patient, whose medical history this information reflects.

Second, the analyst structures the information by determining part-of relations and by linking information to the documents that materialize it (Activity 2.2). For instance, the information medical history is composite: the Patient's personal data and health status are part of medical history. The latter is materialized by a report (document), which is in turn part of the document health record.

The outcome of this phase—led by the requirements analyst—is the *information view*, which is a bridge between the social and the authorization views. For details on this view, see Section 4.3.

6.6   Phase 3: authorization modeling

Permissions and prohibitions are key concepts to analyzing security requirements, for these are used by the actors in a socio-technical system to permit (allow) or prohibit other actors to use (manipulate) their valuable assets. By leveraging the information view of STS-ml, the security engineer defines the permissions and prohibitions that the stakeholders grant to one another.

This phase results in the *authorization view*, fully explained in Section 4.4. This view enables expressing fine-grained relationships concerning *who* can use documents that represent specific information, for *what* purpose, and *how*.

As an example, consider Alice's need to provide to Red Cross BTC her health record when filing an application as a donor. As such, Red Cross BTC should be authorized to access such document. However, Alice may authorize Red Cross BTC to read her health status information to approve her as a donor, but she may prohibit any modification of the information. Also, her authorization may be limited to an objective (scope): Red Cross BTC can read the information to approve Alice as a donor, but not for any other purposes.

6.7   Phase 4: automated analysis

The three views created in Phases 1–3 are tightly related, as they constitute, together, the overall model of the socio-technical system under design. By being different yet complementary artifacts, these views may include inconsistencies and conflicts between security requirements (of the types that are explained in Chapter 5). The purpose of the automated analysis phase is to discover these issues, which are represented in the analysis results artifact.

If errors are detected, the process steps back to social, information, and authorization modeling, till all errors are fixed. While doing so, the security engineer may involve the requirements analyst and stakeholders to negotiate on conflicts that are not resolvable by analyzing the models.

Specifically, automated analysis in STS includes verifying the well-formedness of the model (Activity 4.1, described in Section 5.1), verifying the satisfaction of stakeholders' security needs in the model of the system-to-be (Activity 4.2, explained in Section 5.2), and verifying how threats affect stakeholders' assets (Activity 4.3, detailed in Section 5.3).

The process represented in Figure 6.2 suggests a sequence of analysis activities, but this is also not prescriptive. As a matter of fact, only well-formedness analysis is required to be executed first, for the other two analyses require a well-formed model. As far as security analysis and threat analysis are concerned, they could be performed in any order, as the security engineer chooses to execute them.

6.8   Phase 5: specification

The process followed by the STS method terminates with the specification phase, led by the requirements analyst, which takes an error-free STS-ml model and returns a security requirements specification. This document contains a specification of the security requirements for the socio-technical system under design, while describing the overall system, its stakeholders, details on the created views (social, information, and authorization), and the relevant unresolved security issues.

The activities of this phase are automated by the STS-Tool, which is described in Chapter 7. This is extremely useful; otherwise it would be a tedious and error-prone activity, not to mention unfeasible to conduct manually over growing (in size) goal models at every iteration.

Following the STS method, the specification phase may lead to further iterations. Various versions of the specification can be produced by the requirements analyst and used for discussing with the stakeholders, to improve the model of the socio-technical system based on their feedback.

Notice that, in practice, it may be hard to solve all conflicts [16], and the security requirements engineer may decide to stick with a specification that does not meet (satisfy) all the security requirements, because of unresolvable conflicts or a too high cost to comply with all requirements.

6.9   Chapter summary

The STS method is a security requirements engineering method for the design of secure socio-technical systems. By employing the STS-ml language, the STS method analyzes the problem domain in terms of both social and technical aspects, and ascribes security requirements to the interactions among the actors in the socio-technical system.

This chapter has outlined the method in terms of the main modeling and analysis activities that guide the requirements analysts and security engineers in specifying secure socio-technical systems. This chapter serves as a foundation for applying the STS method in practice, which is illustrated using two industrial case studies from different domains in Chapter 8.

6.10   Exercises

Review questions

Q6.1. What steps are prescriptive in the STS method process?

Q6.2. Can the specification phase be executed before analysis activities are executed? Why (not)?

Q6.3. What are the main roles in the STS method? Can some of them be played by the same individual?

Q6.4. What is the relationship between the STS method and software/system engineering methods?

Q6.5. What are the main artifacts that are used in the STS method?

Q6.6. How is risk analysis embedded within the STS method?

Problems

P6.1. Consider the healthcare scenario of Section 1.5, and the models that you created in problems P4.1 and P5.1. Answer the following questions:

a. Describe the modeling and analysis steps you have followed to build the partial models in the previous chapters. Can you identify any deviation from the STS method?

b. Focus on at least three actors. Follow the steps of the STS method to draw the models. How does the method help you in creating these models? How many iterations did it take you to complete the model?

c. Present the security requirements specification for the said scenario. Is it consistent?

IV STS IN PRACTICE: TOOL AND CASE STUDIES

7 STS-Tool

STS-Tool is a modeling and analysis software tool that supports the STS method described in Chapter 6. The tool fully supports modeling a socio-technical system using STS-ml constructs and the corresponding set of security requirements types (described in Chapter 3), as well as deriving the list of security requirements once the modeling activities are completed. Furthermore, STS-Tool automates the analysis techniques described in Chapter 5, analyses that enable verifying (i) well-formedness, (ii) violation of security requirements, and (iii) threats' effects on actors' assets.

The STS-Tool is distributed as a compressed archive for multiple platforms, and it is free to download from the STS-Tool website: http://www.sts-tool.eu. It comes with a self-update feature that installs bug fixes and new features without the need to manually download new versions. Here we provide details about STS-Tool's modeling features, the automated analysis support, and the security requirements derivation. A detailed user tutorial can be found at http://www.sts-tool.eu/manuals, while complete scenario examples can be found at http://www.sts-tool.eu/tutorials.

7.1 Modeling features

STS-Tool enables the comprehensive modeling of secure socio-technical systems with STS-ml. The tool, a screenshot of which is shown in Figure 7.1, has the following modeling features:

1. *Specification of projects*: STS-ml models are created in project containers. A project refers to a certain scenario and may contain a set of models. Projects can be created, renamed, imported, and exported.

2. *Project explorer*: STS-Tool customizes the Eclipse RNF (Resource Navigator Framework) in such a way that the user can manage files better and organize them into folders and projects.

3. *Diagrammatic modeling*: the tool enables the drawing of diagrams (models). Diagrams are created within projects, and typical operations are supported, such as create, modify, save, load, undo, and redo.

4. *Multi-view modeling*: the tool provides different tabs for the user to model the three supported views of a socio-technical system diagram: *social*, *information*, and *authorization*. Each tab/view shows specific elements and hides others, while keeping always those elements that

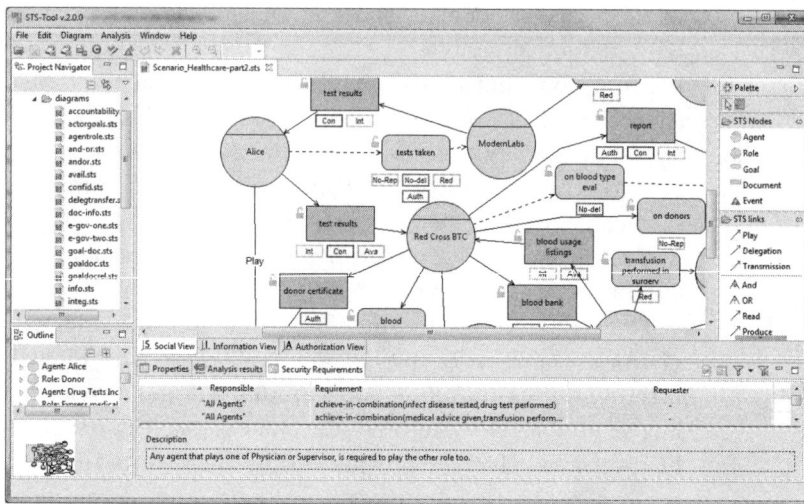

Figure 7.1

Overview of the graphical user interface of STS-Tool.

serve as connection points between the models (e.g., roles and agents). Users can switch from one view to another at any moment. The palette of concepts and relationships is tailored (customized) for each view, grouping the concepts and relationships one needs for modeling in that particular view. Inter-model consistency is ensured by the tool; for example, if the user deletes an agent from the social view, the agent will be removed from the other views as well.

5. *Visual scalability*: to cope with the complexity of STS-ml models, STS-Tool allows users to represent actors either collapsed (only the role/agent shape is shown) or expanded (the whole actor model is shown, that is, the actor together with its goals, documents, and intentional relationships). It is also possible to hide specific elements from one or more views, when they are relevant only for some views. For instance, an actor that does not own any information, have any documents, or manipulate any document might be hidden (omitted) from the information view, for it does not add any details to that view.

6. *Diagram export to different file formats*: STS-ml models (or parts of models, that is, specific elements) can be exported to various formats, such as JPG, PNG, PDF, SVG, and more.

7.2   Analysis support

STS-Tool supports all the analysis techniques that are described in Chapter 5. Specifically, the tool enables

1. *Automated reasoning* for the techniques presented in Chapter 5. While well-formedness checking and threat propagation are implemented in Java, the security analysis verification (Section 5.2) is developed using a Datalog engine. Well-formedness is performed in two steps, depending on the computational complexity of the check, in such a way that automated checks do not slow down or block user modeling:

   - *Online*: the tool performs simple checks on the models on-the-fly while the model is being drawn. For example, the tool does not allow drawing delegations leading to delegation cycles and part-of cycles.

   - *Offline*: more time-consuming checks are performed upon explicit user request (e.g., authorization duplicates), when the user clicks on the dedicated *well-formedness analysis* button.

   Security analysis and *threat analysis* are performed only upon explicit request of the end user (the security requirements engineer).

2. *Visualization of analysis results*: the tool visualizes the results of the analyses and provides explanations for the identified problems. Results are enumerated in a tabular form below the diagram (Analysis result tab, bottom of Figure 7.2) and rendered visible on the STS-ml model itself when selected. A textual description provides further details on the selected analysis result.

   a. *Security analysis* returns the list of security requirements conflicts and violations of security requirements. In Figure 7.2, security analysis has been executed over the healthcare scenario (Section 1.5) and shows some of the detected violations of security requirements. One of the violations, selected in the Analysis results tab, regards non-redelegation. As the description of the selected violation explains, this occurs because Alice has expressed a non-redelegation security requirement over the delegation of goal tests taken to ModernLabs, and the latter delegates goal drug test performed to Drug Tests Inc, which is an and-subgoal of tests taken.

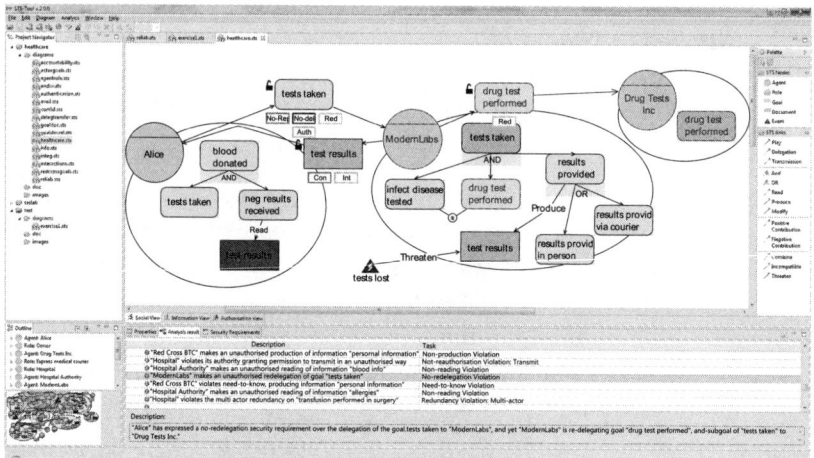

Figure 7.2
Executing security analysis: visualization of results.

b. *Threat analysis* returns the trace of the effects that an event, threatening one or more assets of an actor, has throughout the socio-technical system. The trace contains all concepts and relationships affected by the threatening event and is visualized by highlighting the model. Figure 7.3 shows the impact of event test results lost, threatening ModernLabs' document test results on Alice, who is one of the actors interacting with ModernLabs. The lab transmits to Alice document test results; since this document is threatened, then Alice's goal neg results received is threatened too, as it reads test results. Similarly, goal blood donated is threatened since its and-subgoal (neg results received) is threatened.

7.3   Security requirements derivation

The STS-Tool allows users to automatically derive the security requirements for an STS-ml model using two different outputs:

1. *List of security requirements*: the tool allows the automatic derivation of a list of security requirements expressed as relationships between a *requester* and a *responsible* actor for the satisfaction of a *security requirement*. STS-Tool visually represents these requirements in tabular form, in which they are listed and can be sorted or filtered according

to their different attributes: *responsible*, *requirement*, and *requester* (see Figure 7.4).

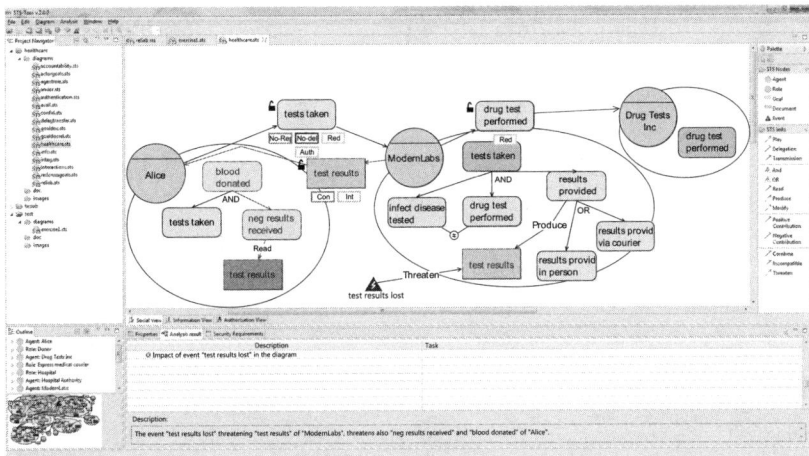

Figure 7.3
Executing threat analysis: visualization of results.

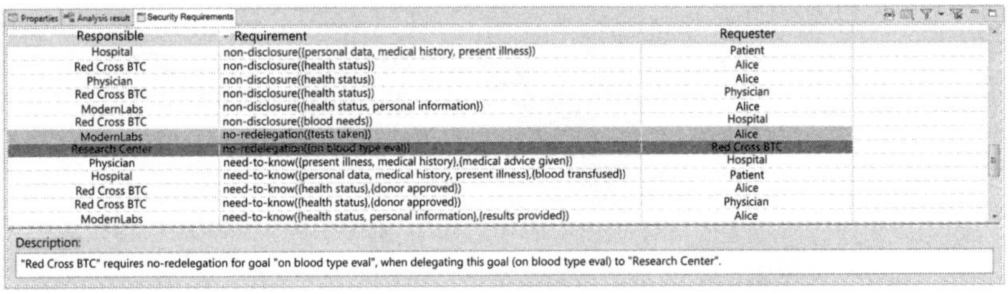

Figure 7.4
STS-Tool: deriving a list of security requirements.

For instance, filtering the requirements with respect to the responsible actor highlights *who* is responsible for fulfilling the security requirements, and what security requirements each actor has to satisfy. On the other hand, filtering security requirements according to their requirement type (as shown in Figure 7.4) groups together security requirements that should be satisfied to fulfill a certain security requirement type. A textual *description* is provided for every *security requirement*, which is displayed below the list of security requirements when one requirement is selected.

2. *Security requirements specification document*: the tool allows the automatic generation of a security requirements document that contains the list of security requirements derived from the model (see Figure 7.5). This powerful feature produces a document that contains comprehensive descriptions of the models; the information can be selected by the user, who can choose to include only a subset of the actors, details on certain views only, and so on. The overall document provides a short description of STS-ml and the STS-Tool, and communicates security requirements by providing details of each STS-ml view, together with their elements. The diagrams are explained in detail providing textual and tabular descriptions of the models.

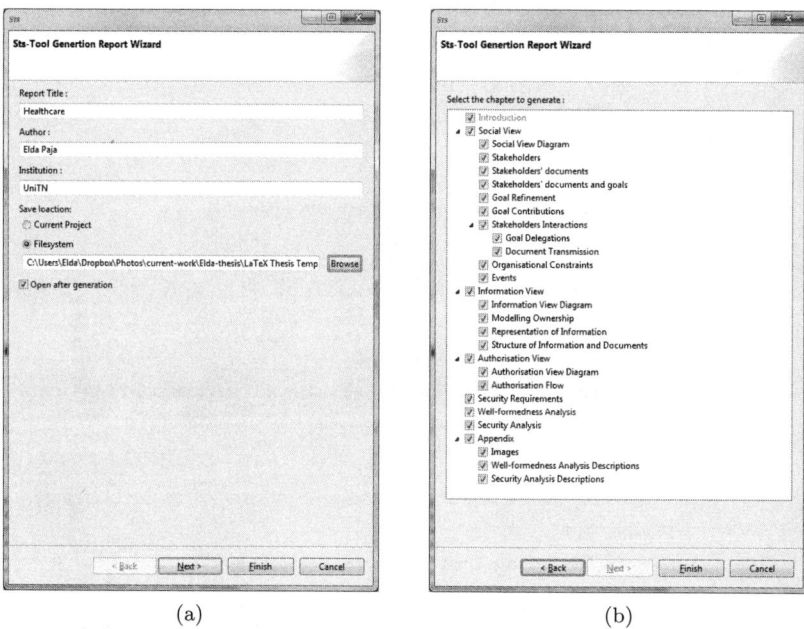

(a)   (b)

Figure 7.5
Customizing the security requirements document.

In generating the document, the user can specify title, author, and organization by editing the fields shown in Figure 7.5a. The document is organized in sections, which the security requirements engineer can decide to include or not in the generated document (see Figure 7.5b). It is good practice to generate the requirements document at the end of the modeling, and after refining the models, to fix eventual errors detected by the automated analyses. This document could be generated at any point of the modeling process, however, and be used as a basis for discussion with the stakeholders to improve the system design (as indicated by the STS method, see Figure 6.1).

7.4   Architectural overview

STS-Tool is a standalone application written in Java, and its core is based on the Eclipse RCP (Rich Client Platform) Framework. STS-Tool is built on top of different frameworks produced by the Eclipse community, as emphasized by the tool architecture presented in Figure 7.6. The layered architecture is composed of three macro blocks, wherein upper layers make use of (rely upon) the layers below.

The bottommost layer—*System Components*—contains the necessary prerequisites for the application: the underlying operating system (Windows, Linux, or OS X), and the virtual machine that executes the Java code.

The second layer—*Eclipse Platform*, also known as Eclipse Rich Client Platform (RCP)—that is well known in the software and systems engineering community, is developed and maintained by the Eclipse Foundation (http://www.eclipse.org). This platform is a powerful framework for building multi-platform standalone applications, just like STS-Tool itself. One of the major advantages of this platform is *modularity*, which is enabled by the use of plugins. Each plugin is an independent module that provides a specific functionality inside the application, and that can be easily added or replaced. Moreover, every plugin can define or consume *extension points*; this feature allows other plugins to contribute functionality to one another. Because of the high modularity of the system, it is possible to add new features with little effort and to maintain code in an easier way. This is very useful for the STS method, as it facilitates the development and integration of extensions.

The third layer—*STS-Tool Components*—includes the most specific modules for the STS-Tool. To support the diagrammatic creation of STS-ml models, a graphical editor is implemented using the Graphical

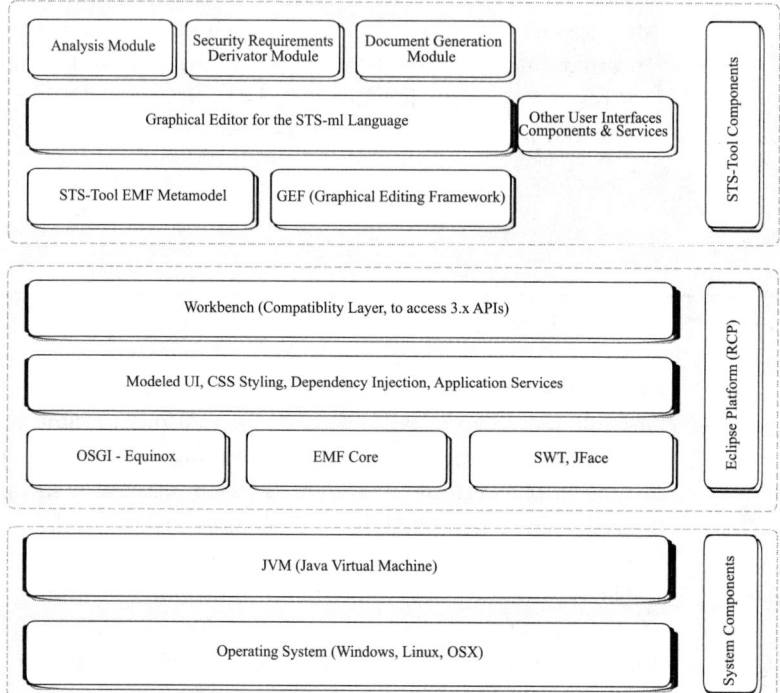

Figure 7.6
STS-Tool architecture: modules.

Editing Framework (GEF) [13]. Specifically, STS-Tool relies on the STS-ml metamodel—specified via the Eclipse Modeling Framework (EMF)—to ensure that the created diagrams adhere to the syntax of STS-ml. The remaining components correspond to the features for executing analysis, deriving security requirements, and generating a security requirements document.

7.5   Chapter summary

The STS-Tool is the modeling tool that accompanies the STS method. It is based on the Eclipse RCP, EMF, and GMF and supports modeling and automated analysis of the created models. A useful feature for practitioners is the generation of a security requirements specification document that feeds the subsequent development phases.

7.6 Exercises

Review questions

Q7.1. Please determine if the following statements are true or false, and explain why.

  a. Roles can be drawn only in the social view.
  b. Information entities can be represented only in the information view.
  c. There can be no redundant authorizations.
  d. The tool allows drawing delegation child cycles.
  e. Security requirements specification documents contain information about security requirements.
  f. Security requirements specification documents can be generated only when security needs are expressed.

Q7.2. Why is the security requirements specification document needed?

Q7.3. Explain the difference between online and offline analysis in STS-Tool.

Q7.4. How does STS-Tool support visual scalability?

Problems

P7.1. Consider the healthcare scenario of Section 1.5. You have built part of this model to address problems P4.1, P5.1, and P6.1. Use the STS-Tool to answer the following questions:

  a. Is the model well formed?
  b. Are there any cases of unauthorized access to donors' sensitive information?
  c. Are there any cases of unauthorized disclosure of donors' medical records?
  d. Are there any unauthorized delegations of rights?
  e. What happens if information on donors' health conditions cannot be confirmed?

Are the answers consistent with your manual reasoning in previous chapters? Describe any discrepancies.

P7.2. Consider the travel agency service (as per problem P3.5) that offers customers the capability of checking various destinations, booking flights and hotels, and so on. Create an STS-ml model (if you haven't done so yet) and run the analysis using STS-Tool.

    a. How does the well-formedness analysis help you? Mention at least two iterations over the models.

    b. Identify three violations of confidentiality requirements.

    c. Are there any violations of availability requirements? What about reliability requirements?

Are the answers consistent with your manual reasoning in previous chapters? Describe any discrepancies.

8 STS Method in Action

This chapter presents the use of the STS method in two case studies from different domains. This serves the purpose of illustrating how the STS method process guides requirements analysts and security engineers (analysts for brevity) in the SRE process, and how the STS-ml language supports modeling secure socio-technical systems from different application areas.

The results of applying the STS method are described step by step for both illustrative domains in Section 8.1 and Section 8.2.

8.1 Trentino as a Lab

Trentino as a Lab (TasLab, http://www.taslab.eu) is an online collaborative platform to foster IT innovation in the Trentino province, Italy [55]. Its aim is to build a community of research institutions, universities, enterprises, and public administration to collaborate on research-intensive IT projects. TasLab provides information on local innovation trends, events, and investment opportunities. It also offers an area where users can match innovation demand (from local government and municipalities) with innovation supply (by enterprises and research institutions), and they can collaboratively write project proposals.

This section presents a scenario that concerns a collaborative project about tax collection that was created using TasLab, the *Taxpayer's Knowledge Base*: http://www.openlivinglabs.eu/livinglab/trentino-lab.

In this project, the innovation demand comes from the local government—the Autonomous Province of Trento (*PAT*)—and the Trentino *Tax Agency*, both requiring a software system that verifies if correct tax revenues are gathered from *Citizens* and *Organizations*. The system is expected to provide a complete profile of taxpayers, to generate reports, and to enable online tax payment.

This is an example of a socio-technical system in which multiple actors interact with one another and with technical systems: citizens and organizations pay taxes online; municipalities (*Municipality*) furnish information about citizens, their addresses, and tax payments; Informatica Trentina (*InfoTN*) is the system contractor; other IT companies develop specific functionalities (e.g., data polishing, development of search modules); the Tax Agency is the system end user; and PAT holds the land register in its database (information about buildings and lots).

These actors exchange and process confidential information. Each actor has its own business policy—that is, goals achieved through processes that manipulate information—and expects others to comply with its security requirements, such as those related to integrity and confidentiality. For example, citizens require that their income tax data be treated confidentially by the municipality and not be disclosed to third parties. Also, PAT is concerned with ensuring that the Tax Agency successfully handles the tax verification process, that is, fulfills the goal of verifying taxes. The tax collection system is not monolithic: its operation depends on successful interaction among taxpayers, the municipality, InfoTN, the Tax Agency, and the TasLab website.

The STS method is applied to the TasLab case study, following its activities concerning modeling, automated analysis, and specification. This section also emphasizes how STS-Tool facilitates these activities through its features.

8.1.1   Social modeling

Following the STS method, modeling starts with the creation of the social view (refer to Figure 6.2). To do so, the requirements analyst has to answer several questions, including the following:

- Which are the stakeholders in the tax collection scenario?
- How can they be represented in terms of roles and agents?
- What goals do they have and how are these goals achieved?
- What documents do actors have and manipulate?
- What social interactions do actors participate in?
- Are there any events threatening stakeholders' assets?

8.1.1.1   Identify stakeholders

As described in Chapter 3, stakeholders in STS-ml are represented via agents and roles. In Figure 8.1, the stakeholders of the TasLab case study are modeled: Informatica Trentina (InfoTN) is an agent, for this company is known to participate in the system since design time, while TN Company Selector is a role, for it is not yet known which company will take over this responsibility, although the responsibilities encapsulated in this role are known. Following the same logic, PAT, BP Engineering Srl, and Okkam Srl are agents, while Tax Agency is a role.

8.1.1.2 Identify assets and interactions

For each actor, *intentional assets—goals—*and *informational assets—documents*[1]*—*are identified, as well as relationships among the intentional elements. For instance, InfoTN wants to achieve goal online system built, for which it has to achieve goals search module built, navigation module built, and system maintained (and-decomposes relationship among these goals); the achievement of the latter goal requires the achievement of goals data completeness ensured and data files stored. InfoTN reads document high quality data to have data completeness ensured, and modifies personal records to achieve goal data refined (see Figure 8.1).

Tax Agency has goal revenue system maintained, which it and-decomposes into goals tax verification performed, tax details verified, data collected, and consultancy offered; goal tax verification performed is further and-decomposed into historic maintained, payers' record created, due taxes calculated, and data completeness ensured; goal data collected requires reading document tax payers' knowledge base; goal payers' record created produces document payers' record, and it is or-decomposed into goals corporate records created and citizens' records created; and, finally, goal due taxes calculated requires reading document high quality data.

STS-Tool support: When first created, roles and agents are visualized along with their rationale (open compartment), so that the analyst can specify the goals or documents (assets) the actors have. The rationales can be hidden or expanded, to give the analyst the option to focus on one or a few roles or agents at a time. Goals have to be placed within the rationale of an actor. STS-Tool facilitates a correct modeling of goal trees by not allowing goal cycles. To such extent, several checks are performed on-the-fly by the tool, such as not permitting the analyst to draw a decomposition link from a subgoal to a higher level goal in a goal tree. Moreover, the tool helps the analyst by allowing goal-document intentional relationships to be drawn only starting from the goal to the document, and not vice versa.

---

1 Remember that a more detailed analysis of informational assets is performed when modeling the information view.

Similarly, the actor models for the other actors are constructed, leading to the outputs shown in Figure 8.1.

Notice that both InfoTN and the Tax Agency do not possess from the beginning all the documents they need to read or modify in order to achieve their goals. For this, they have to rely on other actors. Additionally, not all the goals of InfoTN are its intended goals; some have been delegated to InfoTN from other actors. These elements are obtained through social interactions with other actors, which are supported in STS-ml via social relationships.

An example of *goal delegation* is that of Tax Agency delegating goal data completeness ensured to InfoTN. Moreover, InfoTN delegates goal search module built to TN Company Selector, while it delegates goal cadastre data verified to PAT. An example of *document transmission*, on the other hand, is that of InfoTN transmitting the document high quality data, which it received from TN Company Selector, to Tax Agency. Moreover, PAT transmits document cadastre registry to InfoTN. No knowledge is available about agents adopting the identified roles (thus, there are *play* relationships in Figure 8.1).

STS-Tool support: When drawing a delegation, the tool makes sure that the actor does have at least a goal it wants to pursue, before allowing the analyst to draw the goal delegation relationship starting from the given actor. According to the STS-ml syntax, STS-Tool allows only the delegation of leaf goals, thereby forbidding the delegation of upper level goals. If a leaf goal is delegated and the analyst decides to further decompose this goal within the delegator's model, the tool will prompt him with a message and disallow further decomposition.

Once the goal delegation relationship is drawn, the delegated goal is automatically created within the compartment of the delegatee. Additionally, the tool does not allow the goal to be deleted from the delegatee's compartment unless the delegation is deleted. Importantly, delegation cycles are not permitted by the tool.

8.1.1.3 Express security needs

The analyst analyzes the elicited interactions (goal delegations and document transmissions) and models stakeholders' needs with regard to

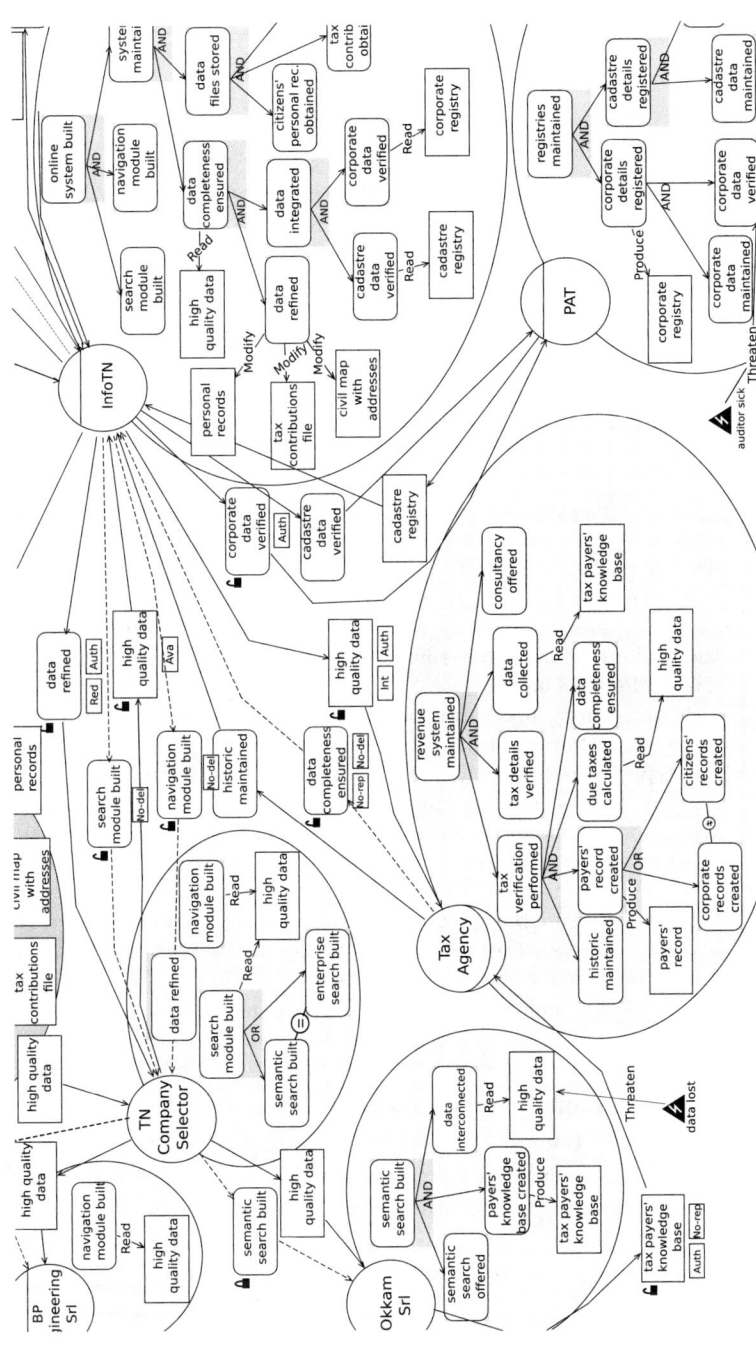

Figure 8.1

Partial STS-ml social view of the tax collection scenario.

security. Additionally, security needs derived from organizational constraints are modeled too.

Over goal delegations. InfoTN expresses several security needs regarding the goal delegations it participates in: InfoTN wants Tax Agency not to repudiate the delegation of goal data completeness ensured; it requires TN Company Selector to ensure true redundancy single for goal data refined; it requires TN Company Selector not to redelegate goal search module built. The delegation of goal data refined from InfoTN to TN Company Selector includes a delegator authentication requirement; the delegation of goal corporate data verified from InfoTN to PAT, on the other hand, includes a delegatee authentication requirement: InfoTN wants the verification to be performed by the same person who maintains the data record.

STS-Tool support: To specify security needs over goal delegations with STS-Tool, the analyst needs to right-click on the delegated goal, for a drop-down list of security needs, and select the desired ones. When at least one security need is selected, a padlock is shown on the affected goal. The selected security need can be shown explicitly by clicking on the padlock, which shows small boxes below the delegated goal; each box has a different label.

More security needs can be specified over the same goal delegation, although certain security needs cannot coexist (for instance, the various types of redundancy, confidentiality, authentication, and so on for all third level security requirement types in Figure 3.16). As such, these requirements cannot be selected simultaneously. The tool facilitates this by allowing the selection of only one security requirement type. The same label is used for the security requirement types of the same category.

Note that availability requirements need a further step, for they require a value (in percentage) in input by prompting users to insert the desired level of availability when this security need is selected. The properties tab can be used to modify the inputted value if needed.

Over document transmissions. In Figure 8.1, Okkam Srl requires Tax Agency not to repudiate accepting transmission of document tax payers' knowledge base; Tax Agency requires InfoTN (the sender) to guarantee the transmission integrity of high quality data; InfoTN (receiver) requires TN Company Selector to ensure an availability level of 70% for the document high quality data; InfoTN wants the transmission of document high quality data to include the receiver's authentication, while Okkam Srl expresses the requirement that the transmission of document tax payers' knowledge base necessitate the sender's authentication.

> *STS-Tool support*: The specification of security needs over document transmissions is similar to that of goal delegations.

Over responsibility uptake. An example of separation of duties is that between goals corporate records created and citizens' records created of Tax Agency, while an example of combination of duties is that between goals semantic search built and enterprise search built of TN Company Selector (see Figure 8.1). There are no examples of separation of duty or combination of duty between roles in this scenario.

> *STS-Tool support*: Note that the specification of security needs over responsibility uptake is different from those over social relationships. In this case, the symmetric relationships *incompatible* and *combine* are used for separation of duties and combination of duties respectively.

8.1.1.4 Model threats

Two events threatening stakeholders' assets are identified: the event auditor sick threatens goal corporate data verified, while the event data lost threatens document high quality data of OkkamSrl (see Figure 8.1).

8.1.2 Information modeling

The modeling activities continue with the second phase, information modeling (see Figure 6.2). To build the information view, a number of questions are considered, including:

- What is the informational content of the documents in the social view?
- Who are the owners of this information?
- What is the structure of information?
- Is there a structure of documents?

8.1.2.1   Identify information/ownership

The analyst should now consider the modeled actors and identify their informational assets, connecting the latter to the actors that own them.[2]. For instance, Citizen is the owner of information personal info, while land details, location, and fiscal code are information entities owned by PAT (see Figure 8.2). Municipality owns information residential address and tax contributions.

Information is represented via documents. For instance, information personal info is made tangible by Citizen's document personal data, while information location is made tangible by document residential buildings.

Information can be represented by one or more documents (through multiple tangible by relationships). For instance, information personal info is made tangible not only by document personal data, but also by document local copy of data of InfoTN, and document personal records of Municipality.

On the other hand, one or more information entities can be made tangible by the same document. For instance, information fiscal code and tax contributions are both made tangible by document corporate registry.

STS-Tool support: STS-Tool allows the relation owns to be drawn starting from the role or agent toward the information it owns, and the relationship tangible by to be drawn only starting from information to documents.

8.1.2.2   Structure information

The information view supports specifying composite information (documents), through the use of part-of relationships, allowing analysts to build a hierarchy of information entities and documents. For instance, this allows representing that information land ownership is part of the information land details, while document land lots is part of document cadastre registry.

---

2 Recall that STS-ml supports shared ownership: more actors can own the same information.

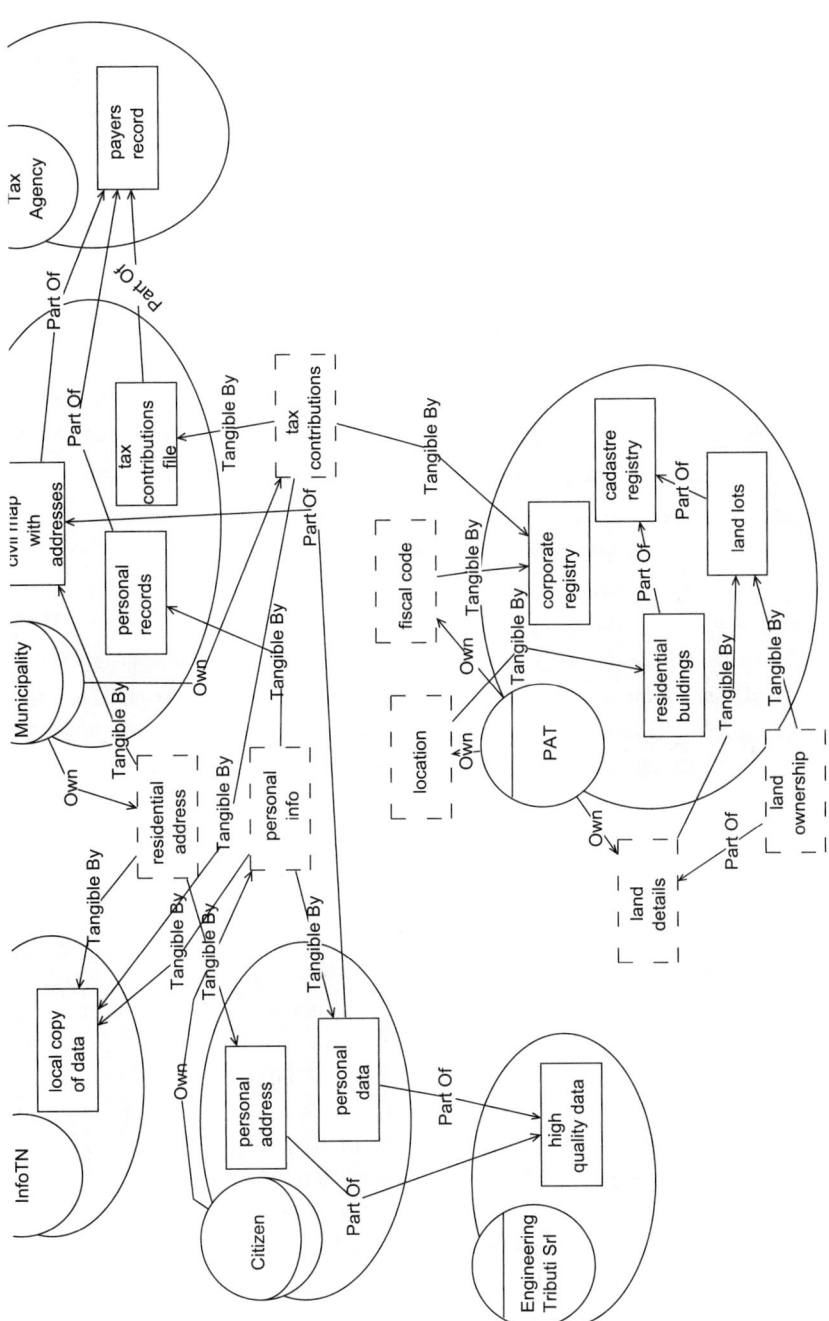

Figure 8.2

Partial STS-ml information view of the tax collection scenario.

> *STS-Tool support*: The tool helps the analyst in building this structure by allowing the **part-of** relations to be drawn only between information or documents. Additionally, cycles of **part-of** are not allowed by the tool.

8.1.3   Authorization modeling

The third phase of the method is authorization modeling (see Figure 6.2). To build the authorization view, the analyst should answer several questions:

- Are there any authorizations granted from the information owners?
- Which are the information entities for which authorization is granted?
- Is authority to transfer authorizations granted?
- Are there any limitations of authority?

8.1.3.1   Model authorizations

The authorization view for TasLab is shown in Figure 8.3. Municipality authorizes InfoTN to read information **personal info**, **tax contributions**, and **residential address**, but it prohibits any modification of such information, in the scope of goal **system maintained**, while granting a transferable authorization (see the dashed arrow).

Authorization modeling includes the specification of security requirements. The authorization from **Tax Agency** to **InfoTN** is an example of a need-to-know security requirement: **personal info**, **residential address**, and **tax contributions** can be modified or produced only for goal **data refined**. There is no example of non-reading in the TasLab case study, for authority to read has not been explicitly prohibited to any actor. Municipality requires the **Tax Agency** not to modify **fiscal code** and **tax contributions**. PAT expresses a non-production requirement on **fiscal codes** to **InfoTN**, by prohibiting the production operation. Non-disclosure is, for instance, required by **Municipality** to **PAT** when authorizing the latter to read information **personal info**, **residential address**, and **tax contributions**, but prohibiting the right to produce and to transmit. An example of explicit non-reauthorization is the authorization from **Citizen** to **Municipality**, since the authorization is non-transferable. Implicit non-reauthorization applies to **Tax Agency**, which has no incoming authorization over information **personal info**.

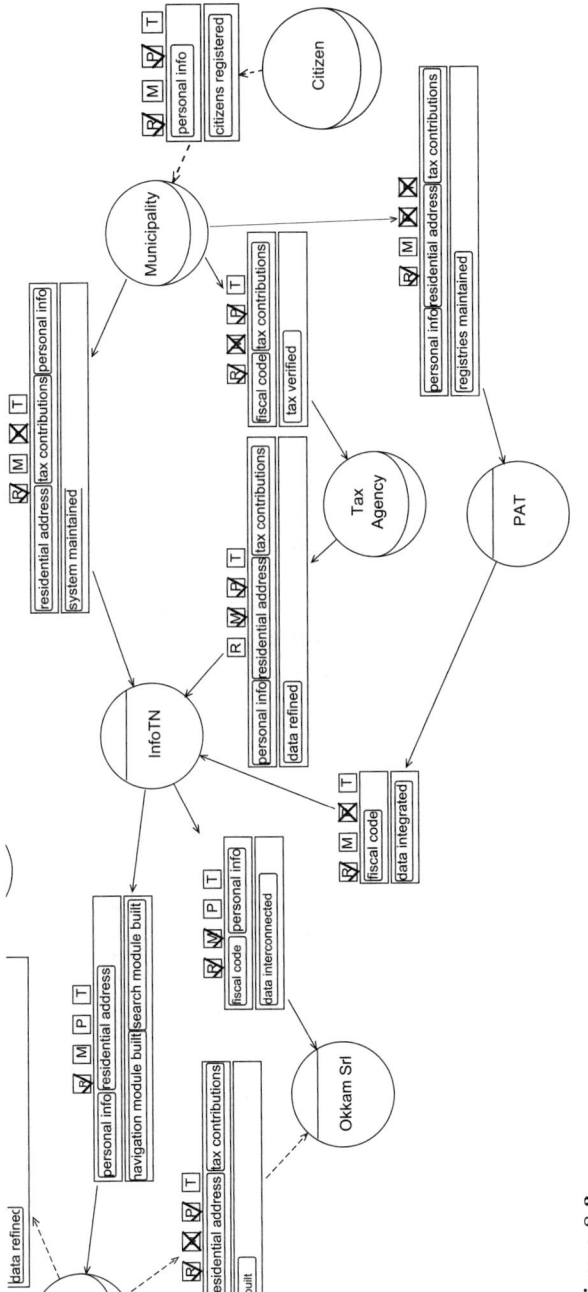

Figure 8.3

Partial STS-ml authorization view of the tax collection scenario.

> *STS-Tool support*: The tool helps the analyst in specifying authorizations, once the relationship has been drawn between two actors, by providing tooltips on specifying permissions and prohibitions, as well as inputting the information by double-clicking the slots representing information and goals.

8.1.4   Automated analysis

The created views can now be checked using the automated analyses supported by STS-Tool, in order to assess well-formedness of the STS-ml model, to identify conflicts, and to propagate the impact of threats over the model.

8.1.4.1   Well-formedness analysis

This analysis finds no errors in the STS-ml model for the TasLab case study. However, this is because only the final models are presented in this chapter. Typically, this technique would identify some problems to be fixed before the analyst could proceed with security and threat analysis.

8.1.4.2   Security analysis

Based on the model that combines the three views: social (Figure 8.1), information (Figure 8.2), and authorization (Figure 8.3), security analysis is executed in order to detect conflicts. This returns a number of issues that are not manifested by just looking at the models:

- *On authority to produce*: Tax Agency authorizes InfoTN to produce documents with information personal info, residential address, and tax contributions to obtain refined data, whereas Municipality authorizes reading only and requires non-production of the same information. This conflict is visualized in Figure 8.4.

- *On authority to modify*: InfoTN grants Okkam Srl the authority to modify documents with information personal info to obtain interconnected data, whereas TN Company Selector requires that no document representing this information be modified.

Improving the model. Conflict resolution activities can be employed by the requirements analysts and security engineers to ultimately reach a consistent model. The former conflict can be resolved by negotiating the provision of adequate rights with the Municipality, while the latter can be fixed by revoking the authorization, given that Okkam Srl does not need it (from the social view).

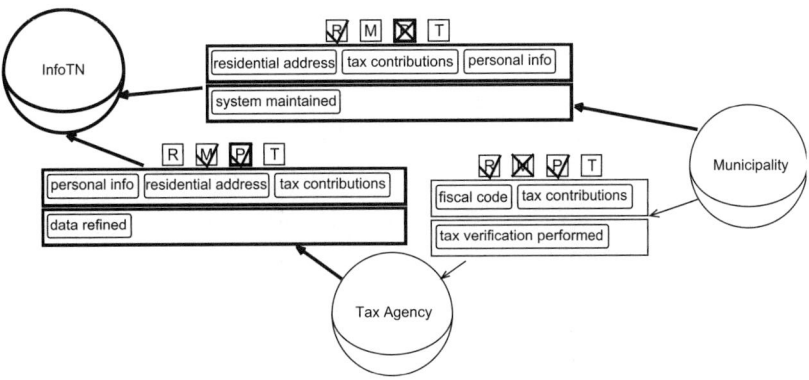

Figure 8.4
Authorization conflict toward InfoTN on authority to produce.

Once authorization conflicts are fixed, the analyst can perform a second round of checks to detect whether conflicts exist between business policies and security requirements, as per Section 5.2.2. In the TasLab scenario, this provides further insights:

- non-redelegation: InfoTN relies on TN Company Selector to refine the obtained data (delegation of data refined). However, Tax Agency relies on InfoTN to ensure data completeness (delegation of data completeness ensured) and requires it not to redelegate this goal. This security requirement is in conflict with the business policy on delegating data refined, since the later is a subgoal of data completeness ensured, for which non-redelegation is required.

- true-multi-red: TN Company Selector has not employed multiple strategies on its own to fulfill the goal data refined, for which InfoTN has required multi-actor true redundancy.

- non-production: PAT makes an unauthorized production of tax contributions, because this information is owned by the Municipality, and the authority to produce is prohibited to PAT.

- non-reauthorized: Tax Agency has no authority to modify information location; nevertheless, Tax Agency authorizes InfoTN to modify location.

- goal-cod: goals semantic search built and enterprise search built should be pursued by the same actor, since a combination of duties requirement is specified between these goals. A conflict occurs because TN Company Selector is not the final performer for both goals (semantic search built is delegated to Okkam Srl).

These conflicts are due to the different policies of the companies. They can be resolved through trade-offs [14] between business policies and security requirements. Notice that relaxation is often not an option, especially if a requirement derives from norms in the legal context.

Improving the model. Through negotiation with the stakeholders, the conflicts are fixed. The non-redelegation conflict can be resolved by relaxing the security requirement, for InfoTN needs to rely on a more specialized actor to refine the obtained data. The true-multi-red conflict can be fixed by removing the delegation from TN Company Selector for goal data refined to Engineering Tribute Srl, but consider or-decomposing it to ensure having more strategies for its achievement. The non-production conflict can be fixed by granting the PAT the authority to produce tax contributions. Similar strategies apply for the other identified conflicts.

8.1.4.3   Threat analysis

The identified threats have to be analyzed in order to decide whether additional security requirements should be specified. Executing threat analysis on the presented models, the following findings are obtained:

- The event data lost threatens document high quality data of Okkam Srl, thereby threatening goal data interconnected, which requires reading this document. In turn, this is a threat for goal semantic search built, for which data interconnected is an and-subgoal; semantic search built has been delegated to Okkam Srl by TN Company Selector, and thus TN Company Selector's goal semantic search built is threatened too. The results are shown in Figure 8.5.

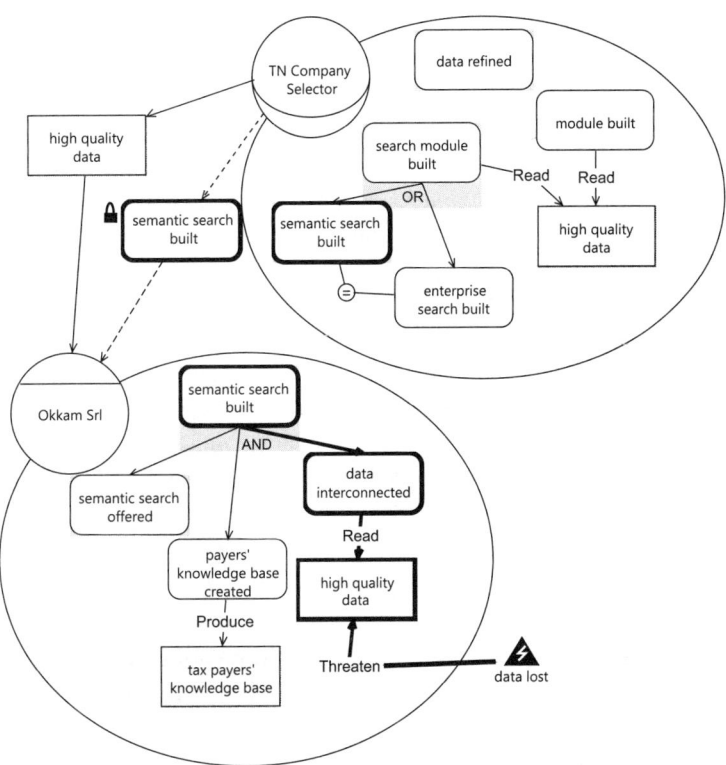

Figure 8.5
TasLab case study: threat analysis results (thicker border) for the event data lost.

- The event auditor sick threatens goal corporate data verified of PAT,
  thereby threatening also this goal's parent corporate details registered.
  In turn, this affects the document corporate registry and its parent reg-
  istries maintained; goal corporate data verified is delegated from InfoTN, and
  therefore InfoTN's goal corporate data verified is threatened; this latter
  goal is a leaf goal in the goal tree (all and-decompositions) of InfoTN,
  and as such it affects goals data integrated, data completeness ensured,
  system maintained, and online system built. Document corporate registry
  produced by PAT is transmitted to InfoTN, and therefore it affects
  InfoTN's document corporate registry too; InfoTN's goal data completeness
  ensured has been delegated to InfoTN by the Tax Agency, and thus, Tax

Agency's goal is threatened too; the effects of this are that goal tax verification performed and revenue system maintained are threatened too.

Improving the model. The analyst is now able to better consider the effects of the events throughout the dense network of interconnections in the socio-technical system. These insights may lead the analyst to choose a strategy in the model that is not influenced by any events, or consider alternative ways for protecting the threatened assets. For instance, in the case of data lost, an alternative could be having enterprise search built to achieve search module built, instead of considering the threatened alternative semantic search built. In this way, no changes to the model are required for its improvement. However, an improvement of the model would be applying a countermeasure, such as making a copy of high quality data and sending it to Okkam Srl in order to deal with the event data lost.

8.1.5   Specification

As a last step, the security requirements specification is generated with the help of STS-Tool, as described in Chapter 7. The outcome is shown in Figure 8.6.

This information helps the analysts further investigate the obtained security requirements. For instance, the analyst can quickly check all the requirements of a certain type, for example, non-disclosure. Figure 8.6 has the derived security requirements ordered with respect to the *requirement* attribute, and highlights one security requirement of type non-disclosure, for which PAT is responsible: *Municipality requires PAT non-disclosure of information personal info, residential address, and tax contributions.*

8.2   E-Government

This section describes a scenario concerning e-Government, which is a variant of the "Land-buying and e-Governance" case study of the Aniketos research project. This case study is reported in Deliverable 6.4 of the Aniketos project, available at `http://www.aniketos.eu/deliverables`.

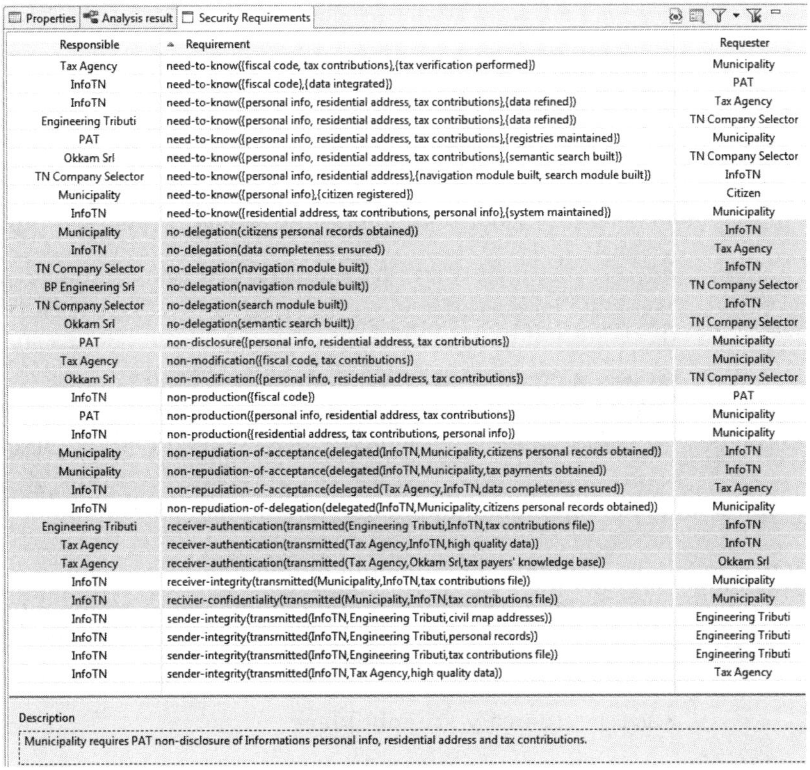

Responsible	Requirement	Requester
Tax Agency	need-to-know({fiscal code, tax contributions},{tax verification performed})	Municipality
InfoTN	need-to-know({fiscal code},{data integrated})	PAT
InfoTN	need-to-know({personal info, residential address, tax contributions},{data refined})	Tax Agency
Engineering Tributi	need-to-know({personal info, residential address, tax contributions},{data refined})	TN Company Selector
PAT	need-to-know({personal info, residential address, tax contributions},{registries maintained})	Municipality
Okkam Srl	need-to-know({personal info, residential address, tax contributions},{semantic search built})	TN Company Selector
TN Company Selector	need-to-know({personal info, residential address},{navigation module built, search module built})	InfoTN
Municipality	need-to-know({personal info},{citizen registered})	Citizen
InfoTN	need-to-know({residential address, tax contributions, personal info},{system maintained})	Municipality
Municipality	no-delegation(citizens personal records obtained))	InfoTN
InfoTN	no-delegation(data completeness ensured))	Tax Agency
TN Company Selector	no-delegation(navigation module built))	InfoTN
BP Engineering Srl	no-delegation(navigation module built))	TN Company Selector
TN Company Selector	no-delegation(search module built))	InfoTN
Okkam Srl	no-delegation(semantic search built))	TN Company Selector
PAT	non-disclosure({personal info, residential address, tax contributions})	Municipality
Tax Agency	non-modification({fiscal code, tax contributions})	Municipality
Okkam Srl	non-modification({personal info, residential address, tax contributions})	TN Company Selector
InfoTN	non-production({fiscal code})	PAT
PAT	non-production({personal info, residential address, tax contributions})	Municipality
InfoTN	non-production({residential address, tax contributions, personal info})	Municipality
Municipality	non-repudiation-of-acceptance(delegated(InfoTN,Municipality,citizens personal records obtained))	InfoTN
Municipality	non-repudiation-of-acceptance(delegated(InfoTN,Municipality,tax payments obtained))	InfoTN
InfoTN	non-repudiation-of-acceptance(delegated(Tax Agency,InfoTN,data completeness ensured))	Tax Agency
InfoTN	non-repudiation-of-delegation(delegated(InfoTN,Municipality,citizens personal records obtained))	Municipality
Engineering Tributi	receiver-authentication(transmitted(Engineering Tributi,InfoTN,tax contributions file))	InfoTN
Tax Agency	receiver-authentication(transmitted(Tax Agency,InfoTN,high quality data))	InfoTN
Tax Agency	receiver-authentication(transmitted(Tax Agency,Okkam Srl,tax payers' knowledge base))	Okkam Srl
InfoTN	receiver-integrity(transmitted(Municipality,InfoTN,tax contributions file))	Municipality
InfoTN	recivier-confidentiality(transmitted(Municipality,InfoTN,tax contributions file))	Municipality
InfoTN	sender-integrity(transmitted(InfoTN,Engineering Tributi,civil map addresses))	Engineering Tributi
InfoTN	sender-integrity(transmitted(InfoTN,Engineering Tributi,personal records))	Engineering Tributi
InfoTN	sender-integrity(transmitted(InfoTN,Engineering Tributi,tax contributions file))	Engineering Tributi
InfoTN	sender-integrity(transmitted(InfoTN,Tax Agency,high quality data))	Tax Agency

Description

Municipality requires PAT non-disclosure of Informations personal info, residential address and tax contributions.

Figure 8.6

List of security requirements for the TasLab case study.

The Department of Urban Planning (DoUP) wants to build an application that integrates the existing back-office system with available commercial services to facilitate the interaction of involved parties when searching for a lot. The *Lot Owner* wants to sell the lot; he defines the lot location and may rely on a Real Estate Agency (*REA*) to sell the lot. *REA* then creates the lot record with all the lot details and has the responsibility to publish the lot record together with additional legal information arising from the current legal framework. *Ministry of Law* publishes the accompanying law on building terms for the lot. The *Interested Party* is searching for a lot and takes the following steps:

1. Accesses the DoUP application to invoke services offered by the REAs.

2. Defines a trustworthiness requirement: only trusted REAs can contact him.

3. Sets a criteria to search and select a *Solicitor* and a *Civil Engineer* (CE) to assess the conditions of the lot.

4. Assigns Solicitor and CE to act on his behalf so that the lot information is available for evaluation.

5. Populates the lot selection for the chosen CE and Solicitor.

Aggregated REA defines the list of trusted sources to be used to search candidate lots; it collects candidate lots from trusted sources and ranks them to visualize the ordered list for the user. The *Chambers* provide the list of creditable professionals (CE and Solicitors).

A detailed report of the application of the STS method was provided for the TasLab case study in Section 8.1; here, a more concise description is given.

8.2.1   Social modeling

The social view for the e-Government scenario is shown in Figure 8.7.

8.2.1.1   Identify stakeholders

The identified roles are Lot Owner, Real Estate Agency, Map Service Provider, Interested Party, Solicitor, CE Chambers, and Solicitor Chambers. The agents are: DoUP Application, Aggregated REA, and Ministry of Law. Recall that Lot Owner and Interested Party are roles, because the actual concrete participants are not yet known; on the other hand, there is only one Aggregated REA and one Ministry of Law in this scenario (thus, these are agents).

8.2.1.2   Identify assets and interactions

The top-level goal of the Lot Owner is lot sold (see Figure 8.8). He could sell the lot either privately or through an agency. Therefore, Lot Owner or-decomposes lot sold into lot sold privately and lot sold via agency. In the second alternative, the Lot Owner interacts with a real estate agency (Real Estate Agency) to create; a lot record should be created, lot information provided, lot location defined, and finally, lot price approved. This is represented through the and-decomposition of goal lot sold via agency into

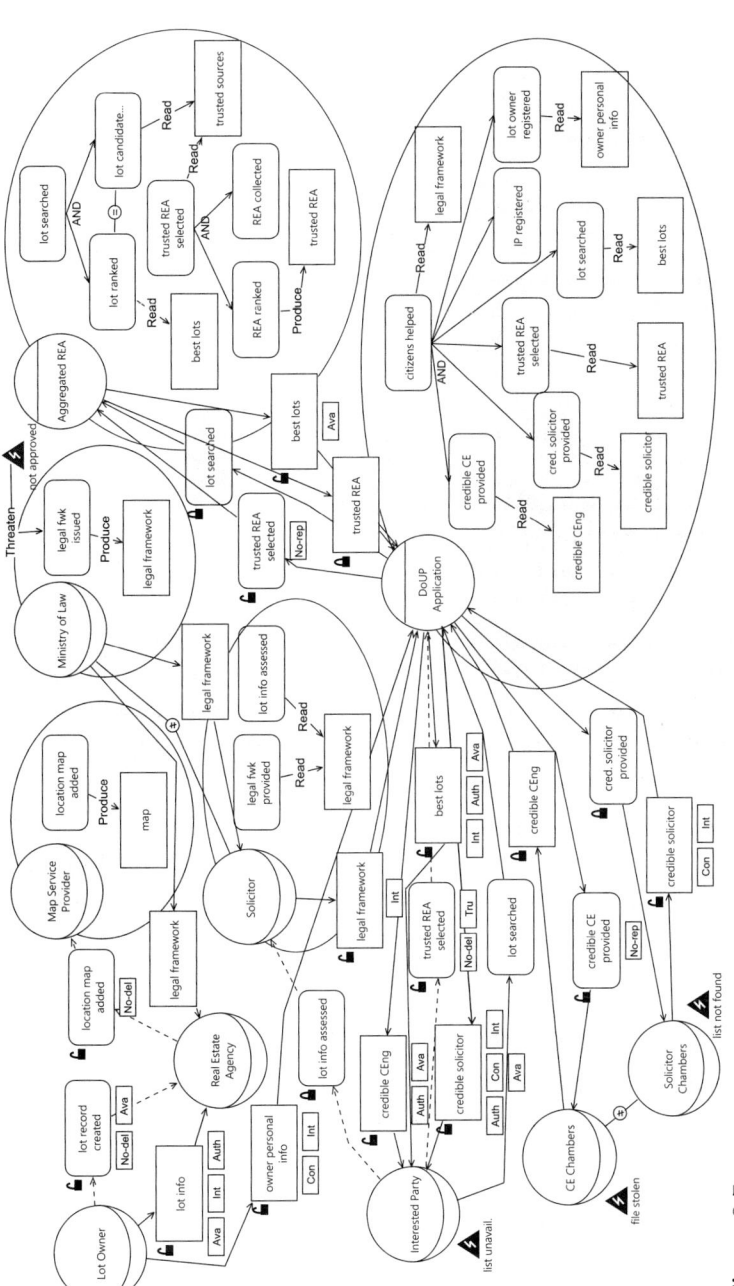

Figure 8.7

E-Government land selling scenario: partial social view.

goals lot record created, lot info provided, lot price approved, and lot location defined.

To actually have the lot sold via agency, Lot Owner delegates goal lot record created to the Real Estate Agency. Consider how the Real Estate Agency achieves its goals: it reads lot info to achieve goal lot record created (the personal data of the owners are necessary to create the lot record). This document (lot info) is produced by the Lot Owner while providing lot information (goal lot info provided). Actors can transmit documents to others only if they possess the required document. For instance, Lot Owner is the creator of lot info (i.e., possesses the document) and transmits this document to Real Estate Agency.

8.2.1.3   Express security needs

Over goal delegations. In Figure 8.8, role Lot Owner requires role Real Estate Agency (REA) not to redelegate the goal lot record created; Interested Party imposes on the DoUP Application a true redundancy single security need for the achievement of goal trusted REA selected; the delegation of goal trusted REA selected from Interested Party to DoUP Application will take place only toward trustworthy application providers; Lot Owner requires the Real Estate Agency to ensure 90% availability for goal lot record created. DoUP Application requires CE Chambers non-repudiation of the acceptance of goal credible CE provided.

Over document transmissions. In Figure 8.9, DoUP Application should ensure sender integrity on the transmission of document best lots to Interested Party; DoUP Application should ensure sender confidentiality on the transmission of document credible solicitor to Interested Party; DoUP Application should ensure an availability level of 94% for the document best lots and an availability level of 90% for the document credible solicitor, when transmitting both these documents to Interested Party.

Over responsibility uptake. As far as separation of duties is concerned, the goals lot record published and location map added are defined as incompatible (not-equal sign, see Figure 8.8). An example of role-SoD is that among roles CE Chambers and Solicitor Chambers in Figure 8.10. With respect to combination of duties, in Figure 8.9, there is a goal-based combination of duties expressed among goals solicitor selected and CE selected of Interested Party.

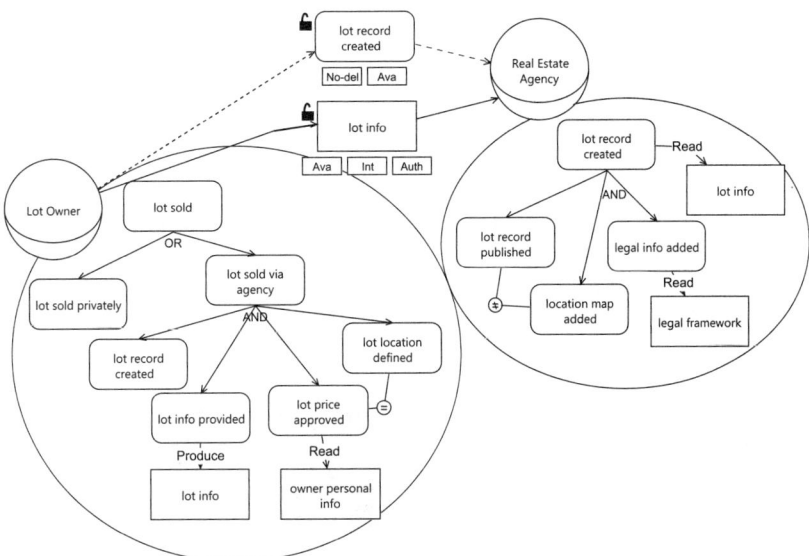

Figure 8.8
Expressing security needs: Real Estate Agency.

8.2.1.4 Modeling threats

Figure 8.7 identifies the events that threaten actors' assets. For instance, event file stolen threatens document credible CE of CE Chambers, while event list not found threatens goal credible solicitor provided of Solicitor Chambers (see also Figure 8.10, which zooms in on part of Figure 8.7).

8.2.2 Information modeling

The information view of the e-Government scenario is shown in Figure 8.11.

8.2.2.1 Identify information/ownership

Lot Owner provides information about the lot: lot info details, which is owned by the Lot Owner himself and is represented (made tangible) by document lot info. Additionally, Lot Owner owns information VAT number, ID card number and lot geo location.

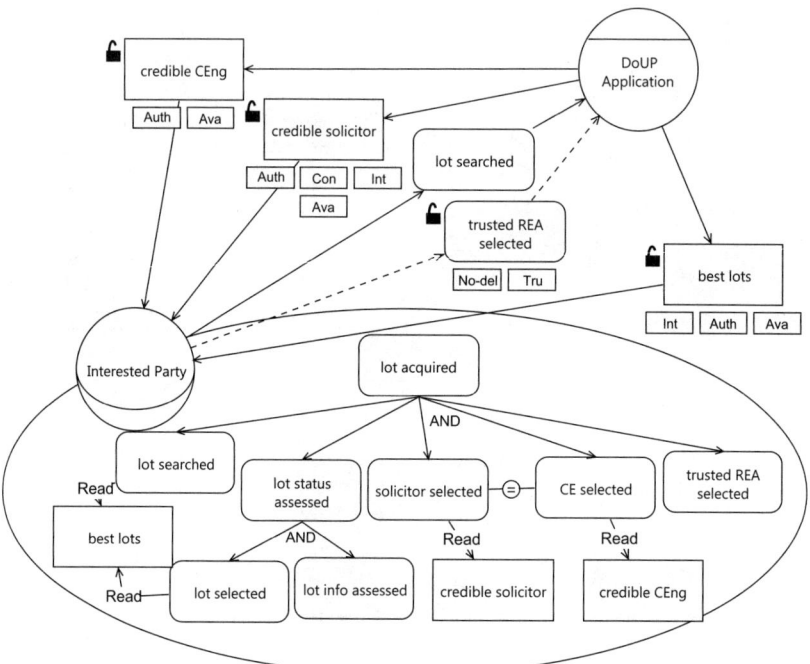

Figure 8.9
Expressing security needs for role Interested Party.

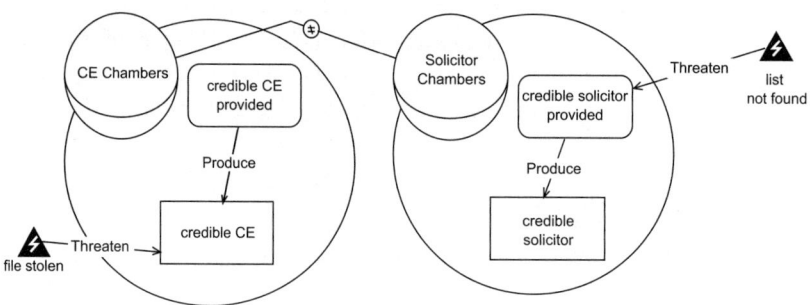

Figure 8.10
Modeling threats.

Aggregated REA owns the information list of credible REA; CE Chambers own the information list of credible CE; Solicitor Chambers own the information list of credible sol; while Ministry of Law owns the information legal info.

Information VAT number and ID card number are made tangible by document owner personal info, while information lot geo location is made tangible by map. Similarly, the rest of the information entities are represented via documents.

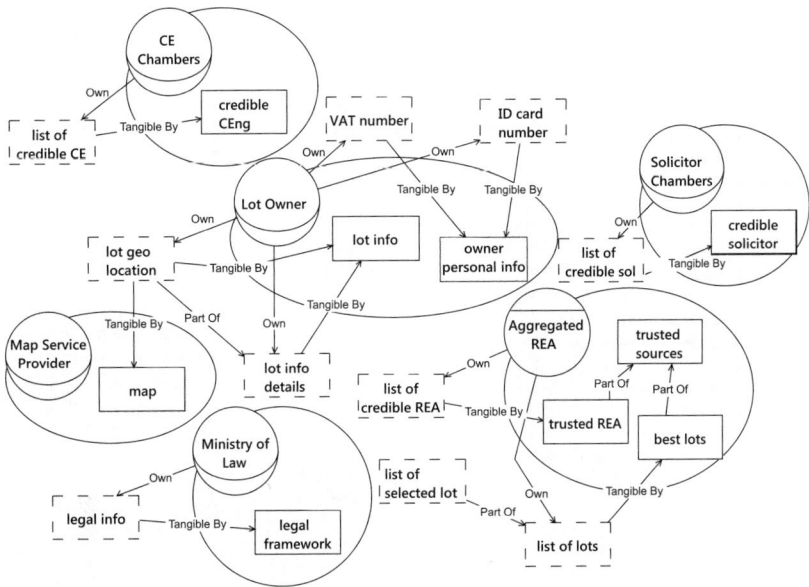

Figure 8.11
E-Government land selling scenario: information view.

8.2.2.2   Structure information

The analyst identifies the part-of relationships among information and documents. For instance, information lot geo location is part of information lot info details, while documents trusted REA and best lots are part of document trusted sources.

8.2.3   Authorization modeling

The authorization view of the e-Government scenario is shown in Figure 8.12.

8.2.3.1   Model authorizations

The Lot Owner authorizes Real Estate Agency to read, produce, and transmit the information lot info details and lot geo location. No prohibitions are specified in this authorization. A prohibition on modifying the information legal info is expressed from the Ministry of Law to Real Estate Agency.

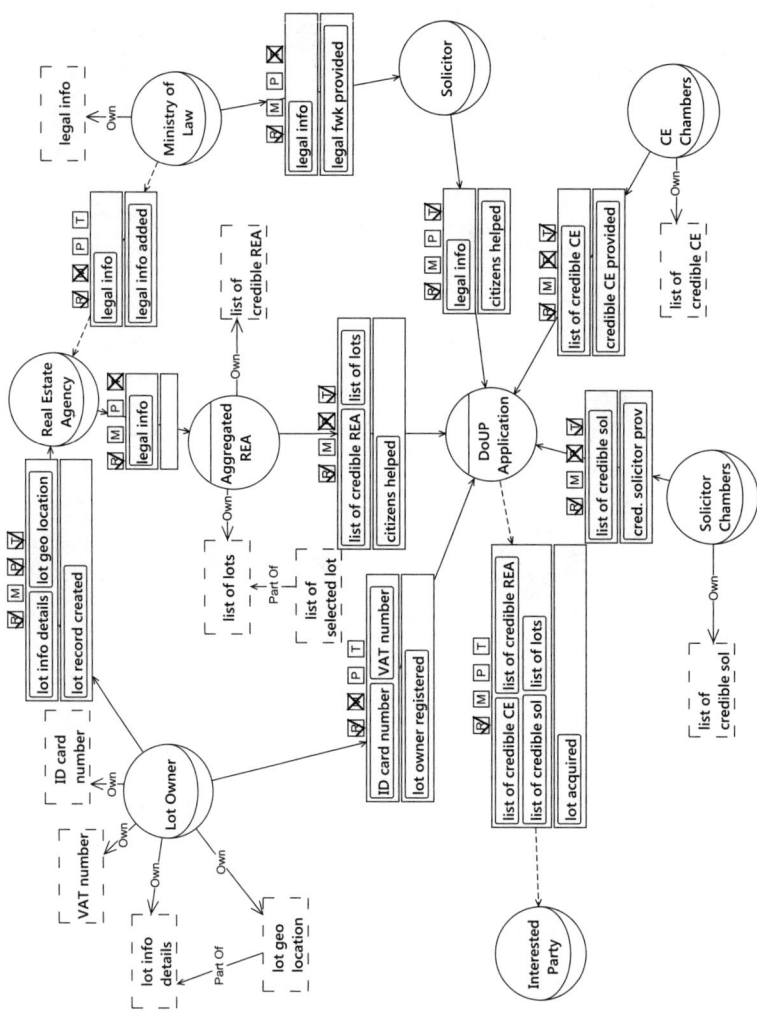

Figure 8.12
E-Government land selling scenario: authorization view.

Moreover, see how the Lot Owner authorizes Real Estate Agency in the scope of goal lot record created, not for every goal of Real Estate Agency. The authorization from Lot Owner to Real Estate Agency is a transferable authorization (continuous arrow line), while the one from DoUP Application to Interested Party granting the authority to read information list of credible CE, list of credible REA, list of credible sol, and list of lots in the scope of goal lot acquired is a non-transferable authorization (dashed arrow line).

Security requirements regarding authorizations are captured by prohibiting operations and limiting the scope. For example:

- Lot Owner authorizes DoUP Application to read ID card number and VAT number only for the purpose of being registered (goal lot owner registered), expressing a need-to-know security requirement to DoUP Application on reading this information only for lot owner registered.

- DoUP Application should not modify documents representing the information ID card number and VAT number, for the authorization from Lot Owner grants the right to read information ID card number and VAT number but prohibits the right to modify these information entities.

- DoUP Application should not produce documents that represent information list of credible solicitors or information list of credible CE, given that the authorizations from Solicitor Chambers and CE Chambers prohibit the operation to produce the respective information entities.

- Solicitor should not transmit documents representing the information legal info, given that the authorization from the Ministry of Law on this information prohibits the right to transmit.

- Non-reauthorization is expressed toward Real Estate Agency concerning the information legal info, for the authorization coming from Ministry of Law on this information is non-transferable (dashed arrow line).

8.2.4   Automated analysis

8.2.4.1   Well-formedness analysis

This analysis does not find any errors when executed over the presented models. Again, it is worth noting that these are final models, which contained problems in their intermediate versions.

8.2.4.2   Security analysis

Security analysis finds several violations of the specified security needs (identifying errors), such as the violation of non-production by the Map Service Provider. As can be seen in Figure 8.12, there is no authorization relationship for Map Service Provider on the information lot geo location (the actor is not even depicted). In Figure 8.7, however, Map Service Provider produces document map that makes information lot geo location tangible (Figure 8.11), whose owner Lot Owner requires non-production of this information (Figure 8.12).

Similarly, there is a possible violation of a combination of duties between the goals lot price approved and lot location defined of Lot Owner (Figure 8.13). A combination of duties requires that the same actor pursue both goals, but there is no single actor achieving both these goals. However, this could change at runtime, and should be verified through

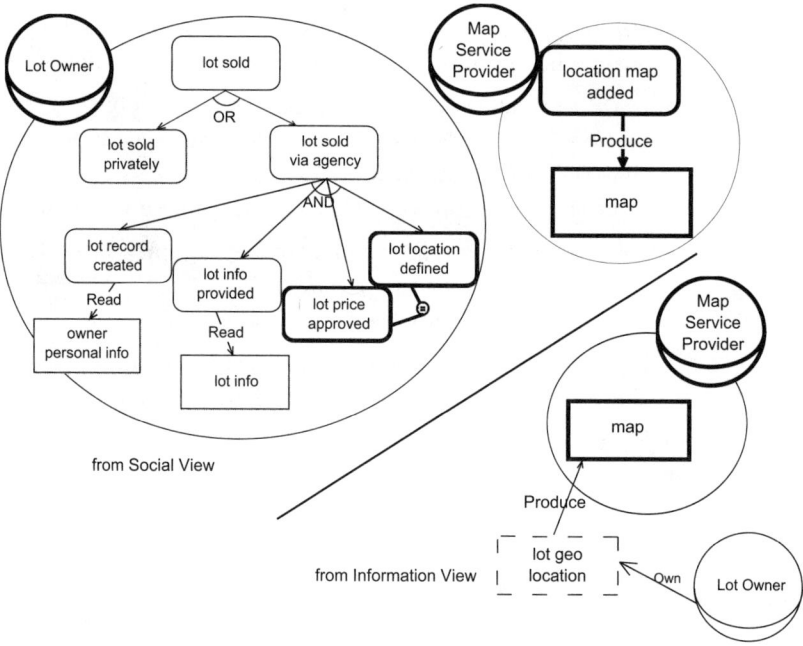

Figure 8.13
Executing security analysis: visualization of results.

monitoring techniques to be deployed in the running system. At the design level, one can verify whether any strategies are undertaken to fulfill the imposed security requirement. Therefore, this conflict is considered a warning, different from the previous one, which is considered an error and needs to be resolved before implementation.

Improving the model. Considering the violations identified by the security analysis, the analyst can resolve them by either relaxing the security requirement or requiring the stakeholder to change its business policy. In the case of the non-production violation, it is clear that the Map Service Provider needs to have permission to produce information lot geo location in order to offer its services. Therefore, negotiation would favor this provider by requiring the Lot Owner to grant it the authority to produce.

As far as the second violation is concerned, the two goals are in fact within the rationale of Lot Owner and could be performed by the same actor adopting this role at runtime. Thus, no changes are made.

8.2.4.3   Threat analysis

Considering the threats shown in Figure 8.10, the analyst runs threat analysis to determine their impact on the overall socio-technical system. The results of this analysis for the event list not found threatening goal credible solicitor provided are shown in Figure 8.14.

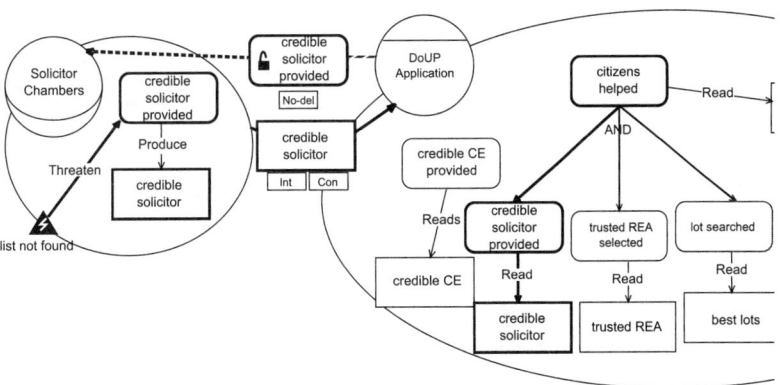

Figure 8.14
Executing threat analysis.

Improving the model. Given the results of the threat analysis, a possible decision would be to consider alternatives to ensuring the provision of credible solicitors rather than accounting on Solicitor Chambers. However, this may not always be possible; it depends on the resources available in the given socio-technical system.

8.2.5   Specification

The security requirements specification is generated with the help of STS-Tool, as described in Chapter 7, and the output is shown in Figure 8.15. In this scenario, the analyst may use this output to answer questions such as *"What are the security requirements for non-modification?"* or *"What about security requirements for non-disclosure of legal information?"*

⬜ Properties ᴸᵉ Analysis ◆ Security Requirements		⊗⊟ ▽ᐧ ⟨ᵏ ⌸ ▢
▲ Responsible	Requirement	Requester
Aggregated REA	non–modification({legal info})	Real Estate Agency
Aggregated REA	non–production({legal info})	Real Estate Agency
Aggregated REA	non–disclosure({legal info})	Real Estate Agency
Aggregated REA	non–repudiation–of–acceptance(delegated(DoUP Application,Aggregated REA,tru	DoUP Application
Aggregated REA	non–repudiation–of–acceptance(delegated(DoUP Application,Aggregated REA,lot	DoUP Application
Aggregated REA	integrity(provided(DoUP Application,Aggregated REA,trusted REA))	DoUP Application
CE Chambers	non–repudiation–of–acceptance(delegated(DoUP Application,CE Chambers,credit	DoUP Application
DoUP Application	need–to–know({ID Card number,VAT number},{lot owner registered})	Low Owner
DoUP Application	non–modification({ID Card number,VAT number})	Low Owner
DoUP Application	non–production({ID Card number,VAT number})	Low Owner
DoUP Application	non–disclosure({ID Card number,VAT number})	Low Owner
DoUP Application	need–to–know({legal info},{citizens helped})	Solicitor
DoUP Application	non–modification({legal info})	Solicitor
DoUP Application	non–production({legal info})	Solicitor
DoUP Application	need–to–know({list of credible CE},{credible CE provided})	CE Chambers
DoUP Application	non–modification({list of credible CE})	CE Chambers

Description:

DoUP Application requires CE Chambers non–repudiation of the delegation of goal credible CE provided, by accepting this delegation.

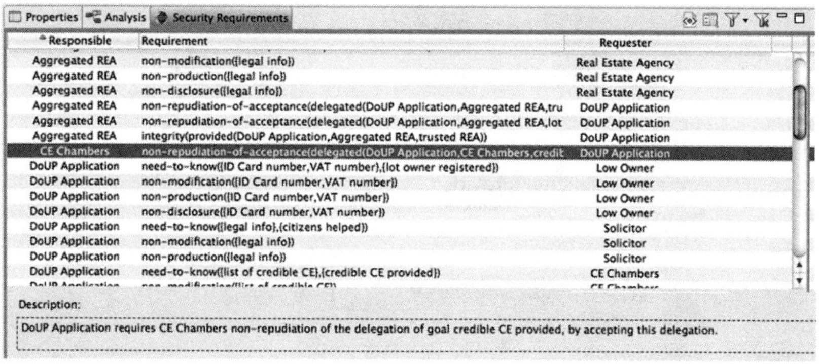

Figure 8.15
Security requirements for the lot-searching scenario.

To identify security requirements for non-modification, the analyst could just order the requirements with respect to the requirement type, to group together requirements on non-modification. Similarly, for the security requirements on non-disclosure of legal information, the analyst would do a similar ordering and look for non-disclosure requirements concerning legal info; only one is found, to be satisfied by Aggregated REA (responsible actor).

8.3   Chapter summary

This chapter has shown two applications of the STS method to case studies, for which STS-ml models have been created (via the corresponding social, information, and authorization views), while capturing security requirements through the modeling of security needs. The obtained models have been analyzed to identify security violations and the impact of threats in the overall socio-technical system.

The chapter has shown how these activities are intended to improve the models (and, consequently, the system design) by resolving the identified conflicts. Finally, the chapter has shown how to derive a security requirements specification for the system under consideration.

8.4   Exercises

Review questions

Q8.1. In Figure 8.1, what is the meaning of the arrow with a "$\neq$" annotation between Tax Agency's goals corporate records created and citizens' records created?

Q8.2. In Figure 8.1, why is the delegation arrow between TN Company Selector and Okkam Srl dashed?

Q8.3. In Figure 8.2, information personal info is owned by Citizen but is made tangible by document personal records, which is in the scope of the Municipality. Is this correct? Why?

Q8.4. In Figure 8.2, within the scope of actor PAT, both documents residential buildings and land lots are part of document cadastre registry. Does this mean that the registry consists of only those two sub-documents? Why?

Q8.5. In Figure 8.3, consider the authorizations for residential address toward InfoTN from actors Tax Agency and Municipality. Are those conflicting? Why?

Q8.6. Consider the information view in Figure 8.11, and look at the tangible by relationships that link information entities lot geo location and lot info details with document lot info. Note the existence of the Part of relationship between the information entities. Doesn't this imply that the tangible by relationship originating from lot geo location is redundant? Why? Can you think of an alternative modeling?

Q8.7. Consider the information legal info in Figure 8.12. Which actors have rights to modify documents containing such information? For what purposes?

Problems

P8.1. Considering the descriptions of the case studies, do you think the presented models are complete? If yes, why? If no, enrich the modeling by considering the remaining actors: (i) draw their models; (ii) draw their interactions (goal delegations and document provisions); (iii) express security needs of their interactions; and (iv) represent threats to their goals and documents.

P8.2. Run analysis on the complete models. What are the findings? How do you interpret them? Can the identified conflicts be resolved? How? Represent the refined models and re-run the analyses.

P8.3. Take the security requirements specification in Figure 8.6. Identify a possible set of security mechanisms for these requirements. If needed, look back at Chapter 2, or consult a book about information and computer security.

P8.4. Repeat the same exercise of problem P8.3 using the specification in Figure 8.15.

V BEYOND THE STS METHOD

9 Alternative and Complementary Approaches

This chapter presents alternative approaches to security requirements engineering and shows how they could be used in conjunction with STS-ml and the STS method, or as a standalone tool to explore certain aspects of security that STS-ml does not cover. These approaches are illustrated using the healthcare scenario that was presented in Section 1.5 and employed in the previous chapters to introduce and explain the concepts of STS-ml.

9.1 Extensions of use cases

Abuse cases [39] extend use cases—part of the Unified Modeling Language (UML) [49]—to capture and analyze security requirements. A *use case* is a list of steps that defines the interactions between an actor (typically a user) and the system under design. On the other hand, an *abuse case* specifies a type of interaction between an actor and the system that leads to results that are harmful for the system. Abuse cases can be used to express a wide range of security concerns, given that the modeling language has no specific primitive.

Figure 9.1 shows an abuse case diagram where actors ModernLabs and Red Cross BTC interact with the socio-technical system in ways that compromise security. Specifically, the diagram shows how ModernLabs may threaten security by altering the results of the tests, losing these results, or publishing on a public website. Similarly, Red Cross BTC may violate

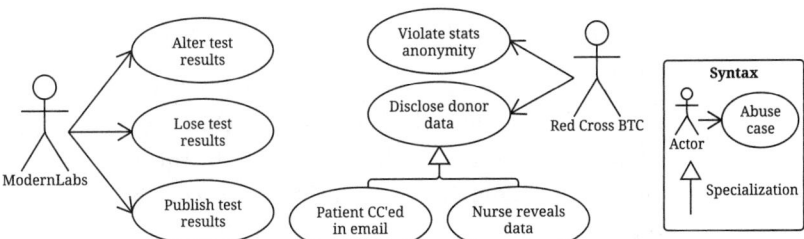

Figure 9.1
Abuse cases: actor-system interactions with harmful effects.

the anonymity of the statistical tests or disclose donor data; the latter case is refined to two more specific ways to disclose data: the patient could be included as a recipient of an email, or a nurse might reveal such data.

Misuse cases [56] take a similar standpoint but employ a richer modeling of negative scenarios. A *misuse case* represents sequences of actions that a *misuser* actor or system performs to cause harm in the system. To denote the negative role of the misuser and the misuse case, these elements are represented in a black background color. Misuse cases include two relationships between use cases and misuse cases: the *threaten* relationship indicates that a misuse case threatens the success of a use case; the *mitigate* relationship specifies that a use case can be used to mitigate a misuse case.

In Figure 9.2, the actor ModernLabs has two use cases: analyzing tests, and transmitting the results of the tests. There are two misusers: a Malicious Employee and a Hacker. The former misuser has the misuse case of

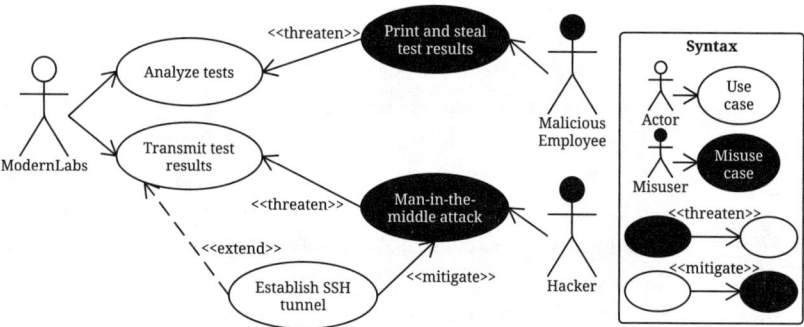

Figure 9.2
Misuse cases, with threaten and mitigate relationships.

printing and stealing the test results, which threatens ModernLabs' use case of analyzing tests. The Hacker has the misuse case of performing a man-in-the-middle attack to threaten ModernLabs' use case of transmitting test results. Here, ModernLabs can employ another use case (transmitting test results by establishing an SSH tunnel) to mitigate the man-in-the-middle attack.

Security use cases [18] are another extension of use cases devised as a means to specify security requirements. They are driven by misuse cases and are based on an analysis of the assets and services to be protected, while considering the security threats from which the said assets and services should be protected. In addition to their graphical notation (illustrated in Figure 9.3), security use cases are enriched with tables that further specify the security use case (see [18] for more details).

In Figure 9.3, ModernLabs has the same two use cases that were presented in Figure 9.2; however, each of them is enriched with two security use cases. The use case Analyze tests leads to security use cases about access control and integrity, in order to tackle the misuse cases of printing test results and modifying them. The use case Transmit test results is

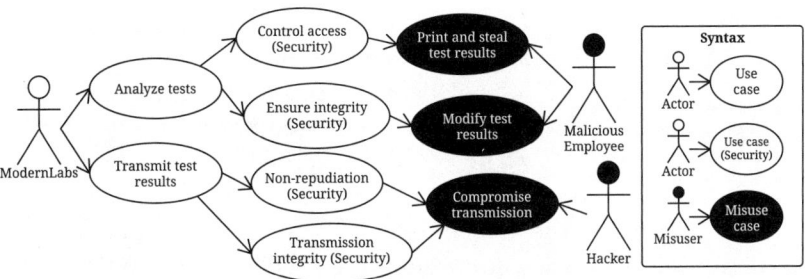

Figure 9.3
Security use cases, in conjunction with use and misuse cases.

connected with two security use cases to guarantee non-repudiation and transmission integrity, both tackling the Hacker's misuse case of compromising the transmission.

Abuse, misuse, and security use cases are a complementary approach to STS-ml. They can be employed as an elicitation tool for identifying what bad actions the actors can execute when using the technical systems within the socio-technical system. Their easy-to-understand graphical notations make them suitable to communicate with stakeholders and to identify the events that may threaten STS-ml assets (goals and information). For example, the identified misuse Modify test results may be turned into an event that threatens the document test results of agent Red Cross BTC in Figure 4.3.

9.2   Anti-goals

Anti-goals [62] abstract abuse and misuse cases to the intentional level; rather than focusing just on what actors do to compromise security, anti-goal models represent *why* actors intend to compromise the security of a system, and *how* they act to achieve their goal. The word "anti-goal" indicates an actor's goal that conflicts with the legitimate/desired goals that the stakeholders ascribe to the system under design.

The proposed method involves the specification of security requirements by incrementally building two models: a model of the system-to-be and an anti-model. The former specifies a set of security goals, making use of specification patterns to elicit candidate security requirements. The anti-model captures how the security goals in the first model could be endangered, deriving the vulnerabilities and capabilities needed to achieve the anti-goals of the security goals. Anti-goals are refined in threat trees, whose leaf nodes represent either vulnerabilities observable by the attacker or anti-requirements implementable by the attacker. The model of the system-to-be is then enriched with new security requirements as countermeasures to be applied to the anti-model.

Figure 9.4 presents an anti-goal model that illustrates part of the anti-goals approach. The model is a hierarchy of goals, denoted as parallelograms, which are related via refinement relationships. The model is constructed starting from goal Test results stolen; this is roughly at the same level of abstraction of an abuse/misuse case. The goal is then refined in two ways:

- Downward, to define alternative ways for the attacker to achieve the anti-goal; either the results printout is stolen, or results are copied to a USB stick. In order to steal the printout, the attacker has to gain access to the data, print the data, and bring away the printout. These are the vulnerabilities for which countermeasures need to be identified;

- Upward, to identify the supposed motivations that drive the attacker's course of action. Here, a possible reason is that data are stolen to sell to competitors; in turn, the ultimate reason could be that the attacker is an insider who would like to make extra money.

Anti-goal models are a very powerful mechanism to investigate security from the perspective of system attackers. Unlike abuse and misuse

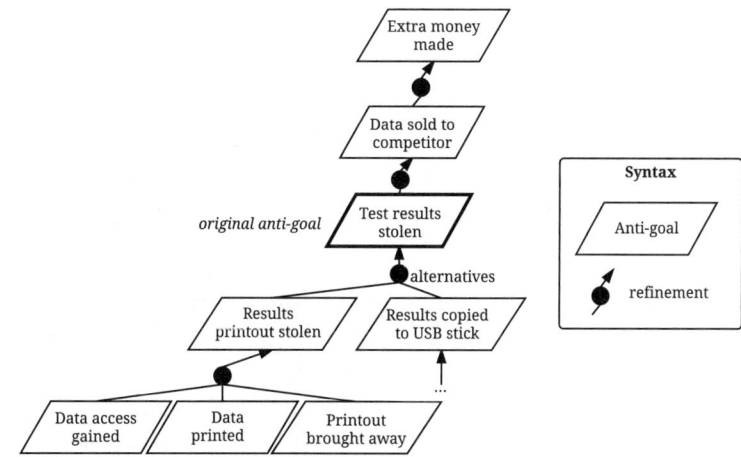

Figure 9.4
Understanding attacker rationale via anti-goal modeling.

cases, the focus is only on intentional attackers and does not include unintentional threats that may occur due to errors or oversight. Anti-goal models can, however, be used as part of threat identification to come up with a more elaborate understanding of the threat. One of the key differences with STS-ml is that, unlike anti-goal models, STS-ml focuses on the social relationships among actors and is not limited to the actor-system interactions.

9.3   Abuse frames

Abuse frames [34] are a modeling and analysis technique that adopts the notation of problem frames [30] and enables the representation of security threats. This technique makes it possible to capture an elaborate understanding of the context where an attacker aims to perform its attack, which includes determining the conditions under which a security violation occurs.

An abuse frame is a composite diagram that consists of multiple elements:

- an anti-requirement that represents a requirement of a malicious attacker
- an asset that is being threatened
- a malicious attacker that aims at attaining the anti-requirement by threatening an asset
- a machine that the malicious actor uses to threaten the asset

These elements are linked by arrows: the *user action* that the malicious user performs on the machine to attain the anti-requirement; the *machine action* that the machine carries out to affect the asset when the user action is completed; and the *effect on asset* that determines when the anti-requirement is met, by result of performing the machine action.

Figure 9.5 illustrates the syntax and presents an example from the motivating scenario (Section 1.5). Such a scenario details a context in which the anti-requirement of the user Malicious Employee is that of obtaining an unauthorized printout of the test results. To do so, the Malicious Employee performs action print requested using the information system to manage the tests; the print request results in the information system sending a print signal on the asset test results. The effect on the asset is that the test results are printed; this way, the Malicious Employee can get out of the building with the printout.

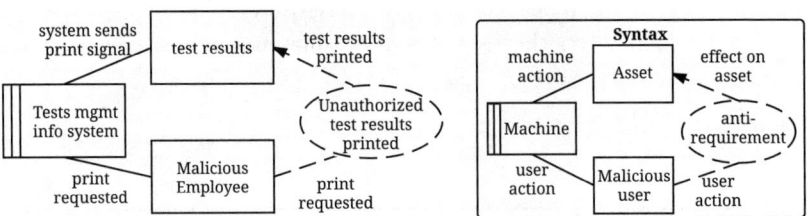

Figure 9.5
Exploiting a machine to achieve a security anti-requirement.

Abuse frames are a powerful technique to represent how malicious users can pose threats and achieve their anti-requirements by using the machine under design, or a different machine. This technique, however, assumes that the machines under design are already specified. STS-ml, instead, looks at the socio-technical context where the machines are not necessarily existent. Therefore, abuse frames are a useful technique for

the design of the machines, after the socio-technical security analysis
with STS-ml has been conducted.

9.4   SecureUML

SecureUML [37] is a modeling language that extends UML designed for
integrating the specification of role-based access control into application
models. As such, its focus is not as comprehensive as that of STS-ml, but
rather it provides a fine-grained language for expressing authorizations.
The basic syntax of SecureUML comprises four types of elements, each
of them defined as a UML stereotype:

- Roles, the active elements that execute actions in the system
- Entities, the passive elements that are being used by roles
- Permissions, association classes between a role and an entity that
  define which methods of the entity a role can call
- Authorization constraints, expressed in the Object Constraint Language (OCL) [64], defining the conditions under which a permission
  is granted

Figure 9.6 illustrates the use of SecureUML for expressing authorizations between roles and entities. In this example, there are two roles
(Operator and Manager) and two entities (TestResults and Patient). The
most interesting entity is TestResults, which features two methods that
enable printing the test results along with patient data or in an anonymous manner. Different permissions are granted to the two roles:

- Managers have the permission managerPrinting for entity TestResults,
  which states that they can execute both types of printing without
  restrictions.
- Operators have the permission operatorPrinting for entity TestResults,
  stating that they can always execute the anonymous printing, while
  the execution of the printing with patient data is granted (see the
  authorization constraint in OCL) only when the operator is currently
  working, and if the operator is in charge of the test results.

Thanks to the use of the precise semantics of OCL, SecureUML models
are constructed in such a way that they can be transformed into an

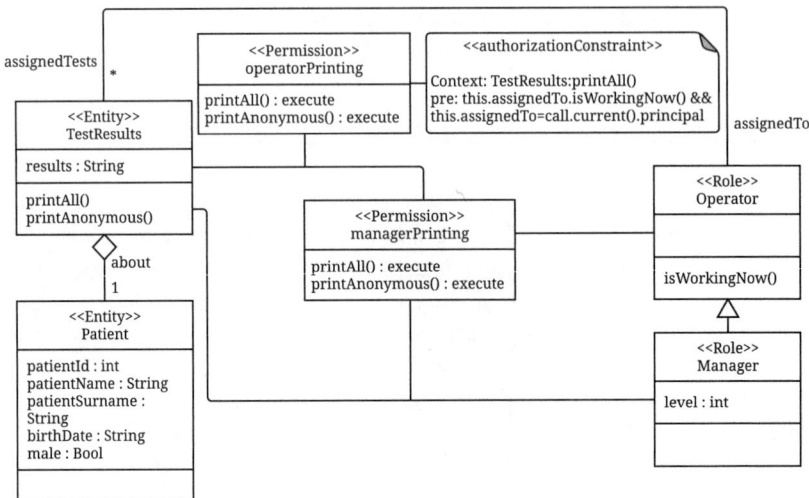

Figure 9.6
Authorizations between roles and entities using SecureUML.

enforceable authorization policy. While positioned within the scope of security requirements engineering, the SecureUML approach targets the very late stages of SRE, after the technical systems have been designed, and the challenge at hand is that of implementing proper mechanisms to protect their operation.

9.5   UMLsec

UMLsec [31] extends UML to develop security-critical systems. The method helps specify security requirements focusing mostly on authenticity, secrecy, and integrity. Security issues are analyzed by representing the behavior of potential attackers (adversaries) and modeling specific types of attackers (stereotypes). Basic security requirements such as integrity are provided/supported via stereotypes and tags, the standard extension mechanism in UML.

Figure 9.7 and Figure 9.8 show a use case diagram and an activity diagram depicting the interactions between Patient and ModernLabs for taking tests and delivering results. The diagrams prevent both actors from cheating, through the use of the stereotype «fair exchange» to the system that contains the use case, and is further explored in more detail in the activity diagram. The tags {start} and {stop} are listed to define that if one of the former actions is executed, then eventually one of the latter will be executed. The property can be automatically verified.

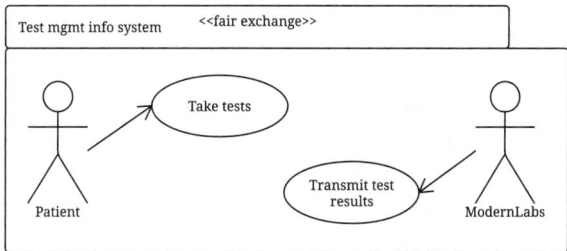

Figure 9.7
Security stereotypes and actions in a UMLsec use case diagram.

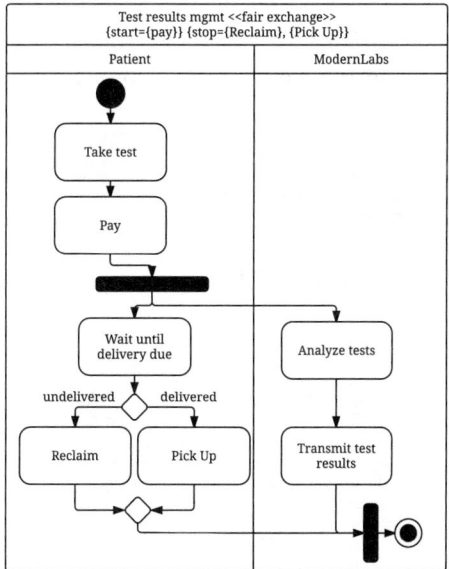

Figure 9.8
Security stereotypes and actions in a UMLsec activity diagram.

UMLsec is complementary to STS. Security requirements expressed in STS-ml can be mapped to UMLsec to pass from a social analysis of security to a technical dimension, in order to ease the transition from the requirements phase to the system design phase.

9.6   Secure Tropos

Secure Tropos [42] builds on the Tropos method [6] and enables modeling security concerns from early requirements to design. As such, Secure Tropos falls within the same class of approaches that STS belongs in, that is, goal-oriented frameworks. Secure Tropos comes with two main security-related modeling concepts:

- *Security constraints*, restrictions related to security issues, which influence the design of the system under development. When such constraints are applied to an actor dependency, such dependency is said to be secure.

- *Secure entities* (goals, tasks, resources), representing elements that are present in the model in order to achieve security constraints.

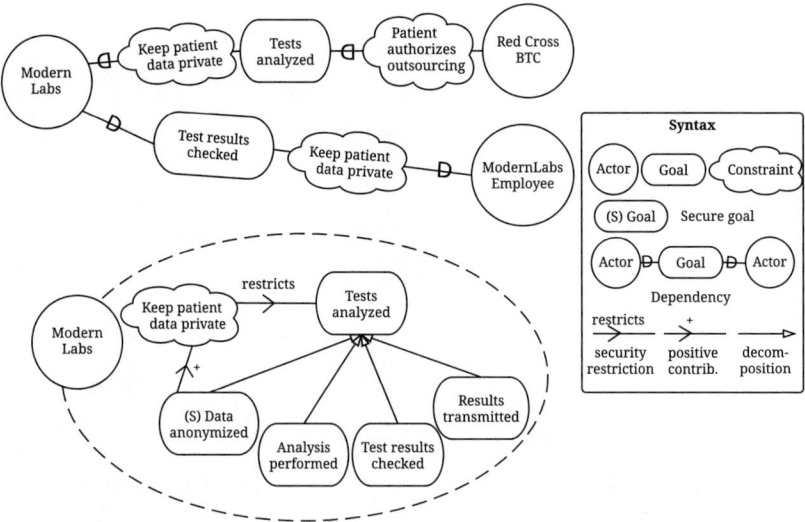

Figure 9.9

Intentional security constraints with Secure Tropos.

Figure 9.9 illustrates Secure Tropos by showing the strategic dependency diagram (top) and the strategic rationale diagram for the actor ModernLabs (below). These diagrams are inheritors of the $i*$ [67] and the Tropos [6] frameworks, and they represent the interactions among the actors in the socio-technical system and the intentional perspective of one single actor.

The strategic dependency diagram shows that Red Cross BTC depends on ModernLabs for goal Tests analyzed, and that ModernLabs depends on its employees to check the correctness of the test results. The diagram also shows security constraints that apply to these dependencies: (i) Red Cross BTC can depend on ModernLabs for analyzing the tests only if the patient has authorized outsourcing; (ii) when fulfilling the dependency for Tests analyzed, ModernLabs has to ensure that patient data be kept private; in turn, (iii) while fulfilling the dependency for Test results checked, ModernLabs Employee has to ensure the privacy of patient data.

The strategic rationale diagram shows some of the intentions of actor ModernLabs. The goal Tests analyzed is decomposed to four subgoals: the first one (Data anonymized) is a security goal, as it helps (positive contribution) achieve the security constraint Keep patient data private, which restricts goal Tests analyzed. The other three subgoals detail how tests are analyzed: the analysis is performed, test results are checked, and results are transmitted.

As already said, Secure Tropos belongs in the same family as STS. The approach that this book proposes is inspired by the notion of security constraint—and, in particular, secure dependency—but has a key difference in that all the supported security requirements have a well-defined meaning that enables automated reasoning. The two approaches could however be used jointly, however; an analyst could start with the informal modeling in Secure Tropos, and later transform the identified security requirements (mainly the security constraints) into the security annotations of STS-ml.

9.7   SI*

SI* [20, 21] is a security requirements engineering framework that relies on organizational concepts, explicitly acknowledging that the system involves the interaction among a number of actors socially depending on one another. It builds on i^* [67] and adds security-related concepts to capture security at the early requirements stage, concepts such as delegation and trust of execution or permission. Being a goal-oriented language, SI* belongs in the same category of modeling languages as STS-ml.

SI* proposes a number of security-related concepts that are used to characterize a socio-technical system. Each of these concepts applies to all of the supported intentional elements: goals, tasks, and resources. For the sake of simplicity, the following explanation is limited to goals, the related security concepts of which are

- *Requires*, denoting that an actor wants a goal to be fulfilled, as such goal is among the actor's objectives

- *Owns*, denoting that an actor is the legitimate owner of the goal and has full authority over it

- *Provides*, representing that an actor has the capability to achieve a goal

- *Delegation of permission*, indicating that one actor delegates to another the right to achieve a goal

- *Delegation of execution*, expressing that one actor appoints another actor to achieve a goal

- *Trust of permission*, representing the expectation of an actor that another actor will not misuse a goal

- *Trust of execution*, denoting the expectation of an actor that another actor is dependable for achieving a goal

- *Distrust of permission*, indicating the expectation of an actor that another actor will misuse a goal

- *Distrust of execution*, expressing the expectation of an actor that another actor is not dependable for achieving a goal

Figure 9.10 illustrates a diagram in SI* and its main primitives. The Donor has goal Blood donated; to achieve such goal, he has to achieve subgoals Test performed, Test analyzed, and Transfusion made. The Donor is not capable of the former subgoal; thus, he requests it and delegates its execution to actor Red Cross BTC. The Donor trusts Red Cross BTC for the execution of goal Test performed; Red Cross BTC provides such goal.

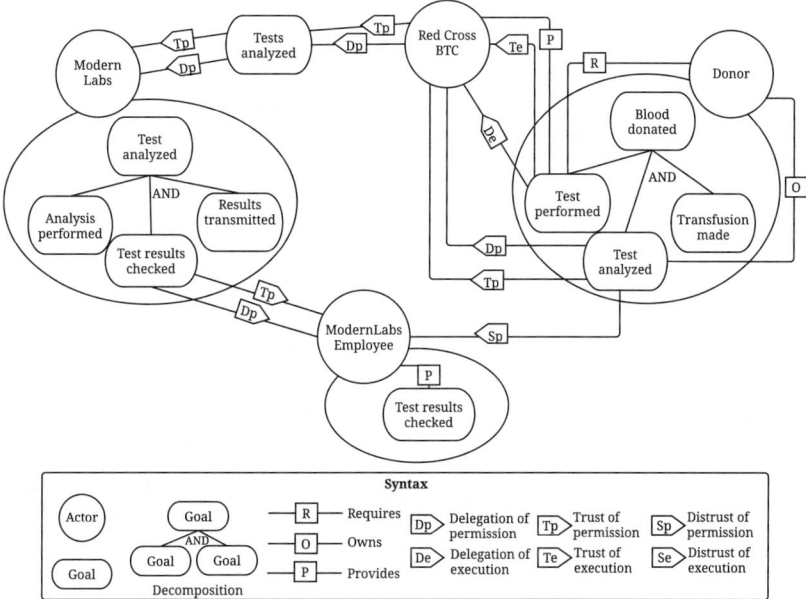

Figure 9.10

Delegations of permission and trust with SI*.

The Donor owns goal Test analyzed, meaning that he determines who can analyze his blood tests. The Donor delegates the permission to Red Cross BTC, and trusts Red Cross BTC that the granted permission will not be misused. In turn, Red Cross BTC delegates this goal to ModernLabs. As part of analyzing the tests, one activity consists of testing the results manually; ModernLabs delegates the permission to do so to ModernLabs Employees. However, the Donor distrusts the ModernLabs Employee for goal Test analyzed (and checking test results is part of such goal).

SI* proposes automated reasoning to check security properties of a model. In Figure 9.10, the distrust relationship from the Donor to ModernLabs Employee results in a violated security property: trust conflict.

SI* could complement STS-ml with the capability to analyze the interplay between the chains of execution, permission, and trust. From a technical standpoint, both STS-ml and SI* rely on a similar formalization in Datalog, making it possible to integrate the features of both approaches.

9.8   SecBPMN

SecBPMN2 [51] is a modeling language for modeling business processes
with security concepts and procedural security policies. It is composed of
two parts: SecBPMN2-ml enables modeling business processes with security aspects, while SecBPMN2-Q supports modeling procedural security
policies. SecBPMN2 extends BPMN 2.0 [44], a well-known standard for
modeling business processes, with 11 security annotations that represent security concepts [8]: accountability, auditability, authenticity, availability, confidentiality, integrity, non-repudiation, privacy, separation of
duties, bind of duties, and non-delegation.

Figure 9.11 shows an example of a SecBPMN2-ml model. It represents two participants, ModernLabs and Red Cross BTC, that interact to
negotiate the price of blood analysis performed by ModernLabs. Each participant has a process that starts with a start event and ends with one or
more end events. Each action performed in the process is represented by

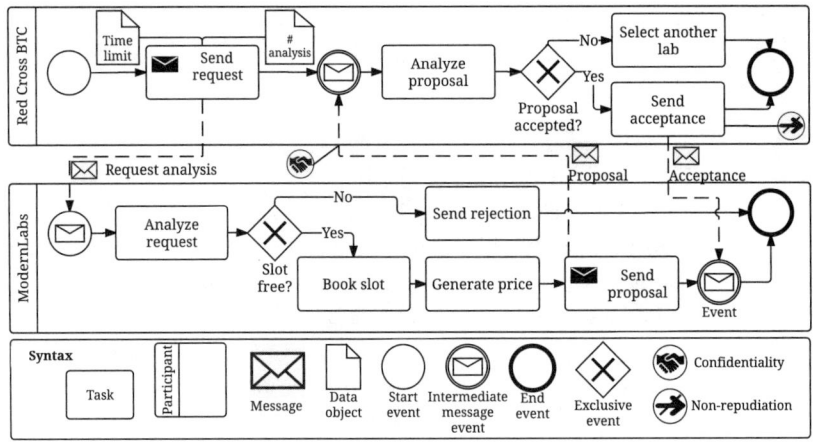

Figure 9.11
Modeling security requirements in business processes using SecBPMN2-ml.

a task element, with solid arrows between tasks representing the control
flow among tasks. Exclusive gateways represent decisions made during
the execution of the process. For example, in Figure 9.11, ModernLabs
executes Analyze request and then checks if there are employees who can

test blood within a certain time limit. Message flows between participants are specified by dashed arrows, while their content is represented
with the message elements. For example, ModernLabs sends a price Proposal to ModernLabs. Documents are specified with data object elements
and linked with a solid line to tasks that use them. For example, the
Time limit data object is used by Send request task. SecBPMN2-ml uses
annotations to specify security concepts; in the example, two security
annotations are used: confidentiality and non-repudiation. The former is
linked to a message flow and specifies that the Proposal message should
not be disclosed to unauthorized users; the latter specifies that Red Cross
BTC should not repudiate the execution of Send acceptance task.

Figure 9.12 shows an example of SecBPMN2-Q security policy. This
policy is verified against all business processes where (i) Send acceptance is executed after Send request (the double pointed arrow between
the two tasks); (ii) the implementation of Send acceptance enforces non-
repudiation; and (iii) the two tasks are executed by Red Cross BTC.

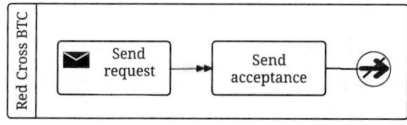

Figure 9.12
Modeling security policies with SecBPMN2-Q.

SecBPMN2 is complementary to STS, as it can be used in later stages
to analyze security in the detailed processes that the participants in a
socio-technical system conduct.

9.9 SQUARE

SQUARE [40] is a nine-step method to elicit, categorize, and prioritize
security requirements for information technology systems and applications. SQUARE does not prescribe specific modeling or analysis techniques, but rather suggests which are the key activities to be conducted.
SQUARE involves the stakeholders of the project as well as requirements
engineers with expertise in security.

The key features of SQUARE are to treat security requirements at the
same time as functional requirements starting with their specification in

the early development stages. The nine steps that constitute SQUARE are as follows:

1. *Agree on definitions*, to guarantee effective and clear communication throughout the process. For example, the notions of non-repudiation, confidentiality, and so on should be agreed on by stakeholders and requirements engineers.
2. *Identify assets and security goals*: What are the assets that the stakeholders want to protect? What are their goals concerning security? What is their priority?
3. *Develop artifacts* that support security requirements definition. SQUARE suggests, for example, system architecture diagrams, use cases, misuse cases, and attack trees.
4. *Perform risk assessment*, identifying threats to the system and its vulnerabilities.
5. *Select the elicitation technique* that fits well with the client organization and project.
6. *Elicit security requirements*: this is the heart phase of SQUARE and includes the execution of the previously selected elicitation technique. This activity results in an initial document with requirements.
7. *Categorize requirements* into essential, non-essential, system, software, and architectural constraints.
8. *Prioritize requirements* to determine those having highest importance.
9. *Inspect requirements* to create accurate and verifiable security requirements.

SQUARE takes a holistic approach to security requirements engineering, starting in the very early phases with a socio-technical analysis to identify the security goals, and ending with the specification of the technical system to be built. The STS method does not focus on the development of a single system, but focuses rather on the design of a socio-technical system. As such, STS could be positioned in steps 2 and 3 of SQUARE, where assets and security goals are created and artifacts are developed, to aid the definition of security requirements. The subsequent activities of SQUARE can be employed to develop individual technical systems that participate in the socio-technical system.

9.10   STRIDE

The STRIDE [26] model, proposed by Microsoft, uses threat modeling to design secure systems by methodologically breaking down the system into components, analyzing each component for susceptibility to threats, and mitigating threats, as a way to discover design problems that potentially allow security breaches and to correct these design-level security problems.

Each of the threats that STRIDE supports leads to a security property/requirement that should be specified to cope with the threat. Table 9.1 summarizes these threats and properties.

Table 9.1

Threats and security properties in STRIDE.

Threat	Sec. property	Description
Spoofing	Authentication	The identity of users is established (or you are willing to accept anonymous users)
Tampering	Integrity	Data and system resources are only changed in appropriate ways by appropriate people
Repudiation	Non-repudiation	Users cannot perform an action and later deny performing it
Information disclosure	Confidentiality	Data are only available to the people intended to access it
Denial of service	Availability	Systems are ready when needed and perform acceptably
Elevation of privilege	Authorization	Users are explicitly allowed or denied access to resources

The STRIDE process is iterative: threats for the system as a whole are analyzed first, then they are broken down into subsystems, and threats for each of those subsystems are analyzed. The process terminates when the analyst is comfortable with any remaining threats.

While successfully employed in large-scale projects, STRIDE does not include any early requirements analysis that studies the socio-technical context. STRIDE starts with the idea of a system already in mind; as such, the requirements that are identified are purely technical, as opposed to the higher level requirements that are captured by STS.

9.11   Security patterns

Security patterns help solve recurring security problems during the design and implementation of systems [54]. A security pattern describes a recurring *security problem* that arises in a specific *security context*, and that presents a well-proven generic scheme for a *security solution*. The intent behind security patterns is to codify security knowledge in a structured and understandable way.

Several types of security patterns exist to address multiple facets of security, including enterprise security, identification and authentication, access control, accounting, and cryptography. A catalog of patterns is built to include both generic patterns and domain- and organization-specific patterns. When studying security, the analyst starts by looking for established solutions from the catalog.

Security patterns, unlike STS-ml, assume that the security goals have already been elicited and analyzed from the stakeholders. Nevertheless, the approach can be jointly used with STS-ml in two ways: (i) security patterns could be defined at the intentional and social level, to describe common solutions to security problems in a socio-technical system; (ii) security patterns can be used to design the individual technical systems that participate in the socio-technical system, after the STS method has been used to determine the security requirements for the socio-technical system.

9.12   Summary

This chapter has reviewed a number of alternative and complementary approaches to the STS method and the STS-ml language for security requirements engineering.

The results emphasize how the landscape of security requirements engineering is broad and many different types of approaches exist, each with strengths and weaknesses. In the following list, the essence of each of these techniques is summarized:

- Extensions of use cases (misuse, abuse, and security use cases) enable representing how a malicious user (or misuser) intends to make use of the system to break the positive use cases.

- Anti-goals extend the study of malicious users to the intentional level, trying to figure out their motivations in order to devise a more effective prevention or response.

- Abuse frames depict the tactic through which a malicious actor makes use of the system/machine to achieve its security anti-requirement.

- SecureUML focuses on the specification of security requirements for a technical system, by studying the interaction among roles and system entities.

- UMLsec extends UML diagrams to support the specification of secure systems.

- Secure Tropos studies security in a socio-technical system using the notions of security constraint and security goal.

- SI* enables modeling and reasoning about chains of delegation, permission, and trust in socio-technical settings.

- SecBPMN supports the modeling of security aspects over business process models as well as the generation and enforcement of security policies.

- The SQUARE method is a comprehensive framework to conduct security requirements engineering; it does not prescribe specific techniques, but rather it defines how to get from the initial analysis needs to prioritized security requirements for a technical system-to-be.

- The STRIDE model proposes a stepwise threat analysis of a system where the system as a whole and its subsystems are examined iteratively.

- Security patterns are used to define a catalog of well-established solutions to recurrent problems in security. These can be used to design a system that effectively embeds best practices in security.

The STS method stands out as a comprehensive tool to study security when designing a socio-technical system. However, alternative approaches from the literature can be employed in conjunction with it.

9.13   Exercises

Review questions

Q9.1. What is the difference between abuse cases and misuse cases?

Q9.2. How do anti-goals relate to STS-ml?

Q9.3. Are abuse frames suitable for modeling the security aspects of socio-technical system? Why?

Q9.4. What are the four main elements of SecureUML?

Q9.5. What are the two main elements of Secure Tropos?

Q9.6. Explain the primitives of SI* that enable representing trust relationships.

Q9.7. What is the difference between the SecBPMN2-ml and the SecBPMN2-Q languages?

Q9.8. Can STS be positioned within the SQUARE method? How?

Q9.9. What is the STRIDE model?

Bibliography

[1] Basin, D., M. Clavel, and M. Egea (2011). A decade of model-driven security. In *Proceedings of the 16th ACM Symposium on Access Control Models And Technologies (SACMAT)*, pp. 1–10.

[2] Baxter, G., and I. Sommerville (2011). Socio-technical systems: from design methods to systems engineering. *Interacting with Computers 23*(1), 4–17.

[3] Becker, M. Y., C. Fournet, and A. D. Gordon (2010). SecPAL: Design and semantics of a decentralized authorization language. *Journal of Computer Security 18*(4), 619–665.

[4] Berendt, B. (2012). More than modelling and hiding: towards a comprehensive view of web mining and privacy. *Data Mining and Knowledge Discovery 24*(3), 697–737.

[5] Blank, R. M., and P. D. Gallagher (2012, September). Guide for conducting risk assessments. Technical Report NIST Special Publication 800-30 Revision 1, National Institute of Standards and Technology.

[6] Bresciani, P., A. Perini, P. Giorgini, F. Giunchiglia, and J. Mylopoulos (2004). Tropos: an agent-oriented software development methodology. *Autonomous Agents and Multi-Agent Systems 8*(3), 203–236.

[7] Cebula, J. L., and L. R. Young (2010). A taxonomy of operational cyber security risks. Technical Report CMU/SEI-2010-TN-028.

[8] Cherdantseva, Y., and J. Hilton (2013). A Reference Model of Information Assurance and Security. In *Proc. of ARES*, pp. 546–555.

[9] Cichonski, P., T. Millar, T. Grance, and K. Scarfone (2012). Computer security incident handling guide. *NIST Special Publication 800*, 61.

[10] Dalpiaz, F., P. Giorgini, and J. Mylopoulos (2013). Adaptive socio-technical systems: a requirements-driven approach. *Requirements Engineering 18*(4), 1–24.

[11] Devanbu, P. T., and S. Stubblebine (2000). Software engineering for security: a roadmap. In *Proceedings of the Conference on the Future of Software Engineering (FOSE)*, pp. 227–239.

[12] Dubois, E., and H. Mouratidis (2010). Security requirements engineering: Past, present and future. *Requirements Engineering 15*(1), 1–5.

[13] Eclipse Foundation (2015). Graphical Editing Framework MVC. Available at www.eclipse.org/gef/gef_mvc/index.php, last checked: September 2015.

[14] Elahi, G., and E. Yu (2007). A goal oriented approach for modeling and analyzing security trade-offs. In *Proceedings of the 26th International Conference on Conceptual Modeling (ER)*, Volume 4801 of *LNCS*, pp. 375–390.

[15] Emery, F. E. (1959). Characteristics of socio-technical systems. Technical Report 527, London: Tavistock Institute.

[16] Ernst, N. A., A. Borgida, J. Mylopoulos, and I. J. Jureta (2012). Agile requirements evolution via paraconsistent reasoning. In *Proceedings of 24th International Conference on Advanced Information Systems Engineering (CAiSE)*, Volume 7328 of *LNCS*, pp. 382–397.

[17] Finkelstein, A., D. Gabbay, A. Hunter, J. Kramer, and B. Nuseibeh (1994). Inconsistency handling in multiperspective specifications. *IEEE Transactions on Software Engineering 20*(8), 569–578.

[18] Firesmith, D. G. (2003). Security use cases. *Journal of Object Technology 2*(3), 53–64.

[19] Giorgini, P., F. Massacci, and J. Mylopoulos (2003). Requirement engineering meets security: A case study on modelling secure electronic transactions by VISA

and Mastercard. In *Proceedings of the 22nd International Conference on Conceptual Modeling (ER)*, Volume 2813 of *LNCS*, pp. 263–276.

[20] Giorgini, P., F. Massacci, J. Mylopoulos, and N. Zannone (2005). Modeling security requirements through ownership, permission and delegation. In *Proceedings of the 13th IEEE International Conference on Requirements Engineering (RE)*, pp. 167–176.

[21] Giorgini, P., F. Massacci, J. Mylopoulos, and N. Zannone (2006). Requirements engineering for trust management: Model, methodology, and reasoning. *International Journal of Information Security 5*, 257–274.

[22] Gollmann, D. (2011). *Computer security* (3d ed.). John Wiley & Sons.

[23] Guizzardi, G. (2006). Agent roles, qua individuals and the counting problem. In *Software Engineering for Multi-Agent Systems IV*, pp. 143–160. Springer.

[24] Gürses, S. (2010). *Multilateral privacy requirements analysis in online social networks*. Ph.D. thesis, Department of Computer Science, KU Leuven, Belgium, May.

[25] Haley, C. B., R. R. Laney, J. D. Moffett, and B. Nuseibeh (2008). Security requirements engineering: A framework for representation and analysis. *IEEE Transactions on Software Engineering 34*(1), 133–153.

[26] Hernan, S., S. Lambert, T. Ostwald, and A. Shostack (2006). Threat modeling-uncover security design flaws using the STRIDE approach. *MSDN Magazine*, 68–75.

[27] ISACA (2012). COBIT 5 for Information Security. Available at www.isaca.org/-cobit/pages/default.aspx?cid=1003566&appeal=pr.

[28] ISO/IEC (2005). Code of practice for information security management (ISO/IEC 27002:2005). Available at www.iso.org/iso/catalogue_detail?csnumber=50297.

[29] ISO/IEC (2011). Information security risk management (ISO/IEC 27005:2011). Available at www.iso.org/iso/catalogue_detail?csnumber=56742.

[30] Jackson, M. (2001). *Problem frames: analysing and structuring software development problems*. Addison-Wesley.

[31] Jürjens, J. (2002). UMLsec: extending UML for secure systems development. In *Proceedings of the 5th International Conference on Model Engineering, Concepts, and Tools*, Volume 2460 of *LNCS*, pp. 412–425.

[32] Kissel, R. (2011). Glossary of key information security terms. Technical Report IR 7298 Rev. 1, National Institute of Standards and Technology.

[33] Laprie, J.-C. (1992). *Dependability: basic concepts and terminology*. Springer.

[34] Lin, L., B. Nuseibeh, D. Ince, and M. Jackson (2004). Using abuse frames to bound the scope of security problems. In *Proceedings of the 12th IEEE International Requirements Engineering Conference (RE)*, pp. 354–355.

[35] Liu, L., E. Yu, and J. Mylopoulos (2002). Analyzing security requirements as relationships among strategic actors. In *Proceedings of the Symposium on Requirements Engineering for Information Security (SREIS)*.

[36] Liu, L., E. Yu, and J. Mylopoulos (2003). Security and privacy requirements analysis within a social setting. In *Proceedings of the 11th IEEE International Conference on Requirements Engineering (RE)*, pp. 151–161.

[37] Lodderstedt, T., D. Basin, and J. Doser (2002). SecureUML: A UML-based modeling language for model-driven security. In *Proceedings of the 5th International Conference on Model Engineering, Concepts, and Tools (UML)*, Volume 2460 of *LNCS*, pp. 426–441.

[38] Masolo, C., L. Vieu, E. Bottazzi, C. Catenacci, R. Ferrario, A. Gangemi, and N. Guarino (2004). Social roles and their descriptions. In *Proceedings of the 9th International Conference on the Principles of Knowledge Representation and Reasoning (KR)*, pp. 267–277.

[39] McDermott, J., and C. Fox (1999). Using abuse case models for security requirements analysis. In *Proceedings of the 15th Annual Computer Security Applications Conference (ACSAC)*, pp. 55–64.

[40] Mead, N. R., E. D. Hough, and T. R. Stehney II (2005). Security quality requirements engineering (SQUARE) methodology. Technical Report CMU/SEI-2005-009.

[41] Mellado, D., C. Blanco, L. E. Sánchez, and E. Fernández-Medina (2010, June). A systematic review of security requirements engineering. *Computer Standards & Interfaces 32*(4), 153–165.

[42] Mouratidis, H., and P. Giorgini (2007). Secure Tropos: a security-oriented extension of the Tropos methodology. *International Journal of Software Engineering and Knowledge Engineering 17*(2), 285–309.

[43] OASIS (2013). eXtensible Access Control Markup Language (XACML) Version 3.0. Available at docs.oasis-open.org/xacml/3.0/xacml-3.0-core-spec-os-en.pdf.

[44] Object Management Group (2011). Business Process Model and Notation 2.0. Available at www.omg.org/spec/BPMN/2.0.

[45] OWASP (2009). Software Assurance Maturity Model. Available at www.opensamm.org/downloads/SAMM-1.0-en_US.pdf.

[46] Pfleeger, C. P., and S. L. Pfleeger (2012). *Analyzing computer security: A threat/vulnerability/countermeasure approach*. Prentice Hall.

[47] PWC (2014a). 2014 Information security breaches survey. Available at www.pwc.co.uk/assets/pdf/cyber-security-2014-technical-report.pdf.

[48] PWC (2014b). Managing cyber risks in an interconnected world. Available at www.pwc.com/gx/en/consulting-services/information-security-survey/download.jhtml.

[49] Rumbaugh, J., I. Jacobson, and G. Booch (2004). *The Unified Modeling Language reference manual*. Pearson Higher Education.

[50] Russell, N., A. H. Ter Hofstede, D. Edmond, and W. M. van der Aalst (2004). Workflow resource patterns. Technical report, BETA Working Paper Series, WP 127, Eindhoven University of Technology.

[51] Salnitri, M., E. Paja, and P. Giorgini (2015). Maintaining secure business processes in light of socio-technical systems' evolution. Technical report, DISI - University of Trento.

[52] Schaeffer, R. C. (2010, April). National information assurance (ia) glossary. Technical Report CNSS 4009, Committee on National Security Systems.

[53] Schneier, B. (1999). Inside risks: Risks of relying on cryptography. *Communications of the ACM 42*(10), 144.

[54] Schumacher, M., E. Fernandez-Buglioni, D. Hybertson, F. Buschmann, and P. Sommerlad (2005). *Security patterns: integrating security and systems engineering*. John Wiley & Sons.

[55] Shvaiko, P., L. Mion, F. Dalpiaz, and G. Angelini (2010). The TasLab portal for collaborative innovation. In *Proceedings of the 16th International Conference on Concurrent Enterprising (ICE)*.

[56] Sindre, G., and A. L. Opdahl (2005). Eliciting security requirements with misuse cases. *Requirements Engineering 10*(1), 34–44.

[57] Sommerville, I., D. Cliff, R. Calinescu, J. Keen, T. Kelly, M. Kwiatkowska, J. Mcdermid, and R. Paige (2012). Large-scale complex IT systems. *Communications of the ACM 55*(7), 71–77.

[58] Stallings, W., and L. V. Brown (2008). *Computer Security.* Prentice Hall.

[59] Stamp, M. (2011). *Information security: principles and practice.* John Wiley & Sons.

[60] Trösterer, S., E. Beck, F. Dalpiaz, E. Paja, P. Giorgini, and M. Tscheligi (2012). Formative user-centered evaluation of security modeling: results from a case study. *International Journal of Secure Software Engineering 3*(1), 1–19.

[61] van Lamsweerde, A. (2001). Goal-oriented requirements engineering: a guided tour. In *Proceedings of the 5th IEEE International Symposium on Requirements Engineering (RE)*, pp. 249–263.

[62] van Lamsweerde, A. (2004). Elaborating security requirements by construction of intentional anti-models. In *Proceedings of the 26th International Conference on Software Engineering*, pp. 148–157.

[63] van Lamsweerde, A. (2009). *Requirements engineering: from system goals to UML models to software specifications.* Wiley.

[64] Warmer, J. B., and A. G. Kleppe (1998). *The Object Constraint Language: precise modeling with UML.* Addison-Wesley Professional.

[65] Welch, D., and S. Lathrop (2003). Wireless security threat taxonomy. In *Proc. of the Information Assurance Workshop*, pp. 76–83. IEEE.

[66] Yu, E. (1995). *Modelling strategic relationships for process reengineering.* Ph.D. thesis, University of Toronto, Canada.

[67] Yu, E. (1997). Towards modelling and reasoning support for early-phase requirements engineering. In *Proceedings of the Third IEEE International Symposium on Requirements Engineering (RE)*, pp. 226–235.

[68] Yu, E., and L. Cysneiros (2002). Designing for privacy and other competing requirements. In *Proceedings of the Symposium on Requirements Engineering for Information Security (SREIS)*, pp. 15–16.

[69] Yu, E., P. Giorgini, N. Maiden, and J. Mylopoulos (2011). *Social modeling for requirements engineering.* MIT Press.

Index